Organic Production and Food Quality

I am very grateful to my dear wife, Moreen, for her encouragement
and support in writing this book and to my late parents, Sam and Mary,
who guided me to the path of learning.

Organic Production and Food Quality

A Down to Earth Analysis

Robert Blair

Faculty of Land and Food Systems
University of British Columbia
Vancouver
Canada

⊛WILEY-BLACKWELL

A John Wiley & Sons, Ltd., Publication

This edition first published 2012 © 2012 by John Wiley & Sons

Wiley-Blackwell is an imprint of John Wiley & Sons, formed by the merger of Wiley's global Scientific, Technical and Medical business with Blackwell Publishing.

Editorial offices: 2121 State Avenue, Ames, Iowa 50014-8300, USA
The Atrium, Southern Gate, Chichester, West Sussex, PO19 8SQ, UK
9600 Garsington Road, Oxford, OX4 2DQ, UK

For details of our global editorial offices, for customer services and for information about how to apply for permission to reuse the copyright material in this book please see our website at www.wiley.com/wiley-blackwell.

Library of Congress Cataloging-in-Publication Data

Blair, Robert, 1933-
  Organic production and food quality : a down to earth analysis / Robert Blair.
    p. cm.
  Includes bibliographical references and index.
  ISBN-13: 978-0-8138-1217-5 (hard cover : alk. paper)
  ISBN-10: 0-8138-1217-8
 1. Natural foods–Processing. 2. Food–Quality. 3. Consumers–Attitudes. I. Title.
  TX369.B53 2011
  363.19′29–dc23

                                                                2011022691

A catalogue record for this book is available from the British Library.

This book is published in the following electronic formats: ePDF [9781119951858]; Wiley Online Library [9781119951889]; ePub [9781119951865]; Mobi [9781119951872]

Set in 9.5/12.5pt Palatino by Thomson Digital, Noida, India
Printed and bound in Malaysia by Vivar Printing Sdn Bhd

1 2012

Contents

Foreword

Organic, local, factory farming, pesticides, nutrition, health, food safety and environmental sustainability are some of the key buzz words surrounding our food supply today. As consumers have moved further away from an understanding of how their food supply is produced, new belief systems have sprung up that exploit the disconnect between the reality of production (and the science behind it) and the perception of consumers. As the world in general becomes more and more complex and individuals have less control over it, many people seek to control their own worlds in some of the few remaining ways that are possible. Choice of food remains one of the few options. As people move to using new information sources such as the internet, often with less "quality control" over content, the quality of the information available decreases. As people move away from communities of common religions and other older traditional values, new values and religions are needed. As people become more affluent, but have less time to spend doing things they enjoy, they look for ways to differentiate their spending patterns by creating "meaningful" expressions of their concerns.

So what is really important to food consumers, and what might provide a solution to these consumer needs, whether real or perceived? It is a challenge that the food industry has to meet. Traditionally, the food industry was also asked to do this at a reasonable cost, but that constraint seems to be loosening as people have more funds for discretionary spending. But as we are also learning, we need to evaluate which "externalities" are actually incorporated into retail pricing and how this complicates the impact of these costs. Who is subsidizing whom?

That brings us back to the first word in this foreword: "organic". Organic foods are proliferating. Although the growth is significant (easier from a small base), it is still a niche market. And the question is whether organic actually provides a mechanism/production method for providing the benefits that consumers "believe" organic food possesses and consumers believe they want. The first attempt at a comprehensive and balanced review of the scientific literature covering the comparison of organic agriculture with other forms of agriculture comes, to the best of my knowledge, in this book by Robert Blair. The news it gives will be viewed differently by different people. For some it fails to be the "slam dunk!" that they want in either direction! (Sorry, but life rarely comes out so one-sided.) Those who strongly support organic agriculture are going to attack Blair and his data for being biased in favor of conventional agriculture – a tool of big business, big farming and capitalism. Those who think organic is a "con" job are probably also going to be disappointed and say he is soft on organic because he really loves the small farmer and his urban farmer's market.

So, what do we make of his book? It sets out clearly, I believe, what we know and – just as importantly, for a scientist – what we don't know. He shows clearly that some perceived differences between different forms of agriculture are real and that these lead to some relevant consequences. He shows that many other differences are not real and that this also leads to some consequences because, in the past, people have made decisions based on the assumption that some of these differences were real. But the bottom line, which comes through clearly, is that the American and European food supplies are safe, nutritious, wholesome and healthy, regardless of production methods. The issues of flavor and freshness may in fact be key features of the current incarnation of organic, but they are actually predicated on variables that are more related to the cultivars selected and the handling and distribution of the food supply than on organic as a production system. So local may have some benefit in the hedonistic areas but these are being lost as organic becomes mainstream and large companies see an opportunity to obtain higher profit margins from organic and join the program.

In the United States, organic is controlled by the government. The rules, whether fair or not, whether consistent with what people really want from organic, whether costly or not, are the ones that organic will need to meet in the future and, as such, they probably slightly favor the larger companies that can afford the supervision charges, that have the resources to participate in the process that decides what the rules are going to be, and that can get organic products widely distributed. It may have been the case that the organic growers, processors and retailers failed to recognize and accept multiple standards that reflected at times genuine differences among the "converted" organic farmers and instead decided to ask the government to step in (as part of the 1990 Farm Bill), thus giving up their "religion" to government control. Moving forward, it is important for researchers to be more rigorous in their methodology and data presentation, and that all those involved in teaching about our food supply need to incorporate more of a balanced view of the conflict between idealism and practicality and to put more emphasis on proven facts. Consumers in the future may also need to think about whether they want to continue to support this expensive method of food production with its religious overtones, without any real "show stopper" advantages, and instead move on to "local", focus on some other issues related to our food supply, or just buy a range of foods that are available today, relax a bit more, just enjoy life and eating, and save their money to buy electric cars, solar water heaters, and personal windmills to generate the electricity they will need in the future. And some might even take public transportation or return to riding a bike. This book suggests that some of these alternative suggestions might do more for the greater good. So whatever you eat, enjoy it.

Professor Joe M. Regenstein
Cornell University

1 The Shift to Organic Food

This book is the first comprehensive text, based on an unbiased assessment of the scientific findings, on how organic production methods influence the quality of foods. In this context "quality" is taken to refer to nutrient content; freshness, taste and related aspects which are obvious to the consumer; also other attributes which are not immediately obvious to the consumer but which are perceived to be associated with organic foods. These are: relative freedom from harmful chemical and pesticide residues, and from hormonal residues; and also "healthfulness" (ability to enhance or promote health in the consumer). In some publications these latter attributes are designated "safety" aspects, but this is not a completely satisfactory designation since it implies that some foods are not safe.

Purchasers of organic foods believe that these products are superior to conventional foods in terms of quality and safety. The available data confirm that there is a growing market for organic foods, if they can be delivered at a price acceptable to the consumer.

The book addresses issues that the food industry and consumers raise about organic food in relation to conventional food, and assesses the relevant scientific findings in the international literature as well as the results of food monitoring programs in North America, Europe and Australia/New Zealand. Documented findings related to the nutritional quality and "healthfulness" of organic food are assessed, as are findings on the motivation of consumers to buy organic food.

Background

The organic system of farming was developed in Europe over 100 years ago by proponents such as Rudolph Steiner in Austria, Albert Howard in the United Kingdom, and Hans-Peter Rusch and Hans Müller who developed "biological agriculture" in Switzerland. The first use of the term "organic farming" appears to have been by Lord Northbourne in the United Kingdom. It derives from his concept of "the farm as organism". He differentiated between what he called "chemical farming" and "organic farming". Sir Albert Howard's concept of soil fertility was centered on building soil humus with an emphasis on a "living bridge" between the soil

Organic Production and Food Quality, First Edition. Robert Blair.
© 2012 John Wiley & Sons, Ltd. Published 2012 by John Wiley & Sons, Ltd.

and the life it contained (such as fungi, mycorrhizae and bacteria), and on how this chain of life from the soil supported the health of crops, livestock and humans. Steiner went on to propose "biodynamic agriculture", a method of organic farming that has its basis in a spiritual view of the world, using approaches such as fermented herbal and mineral preparations as compost additives and field sprays and the use of an astronomical sowing and planting calendar. This farming method became popular in Australia.

Lady Eve Balfour was influenced by the work of Sir Albert Howard and set up the Haughley Experiment on adjacent farms in England in 1939 to compare organic and conventional farming. The experiment was taken over by the Soil Association in 1947, which for the next 25 years directed and sponsored it. The work had a design flaw – no replication – which probably explains why the results were never published in any scientific journal. Based on the early findings, Balfour published a book, *The Living Soil*, in 1943, which did not receive a good review in the journal *Soil Science*: "The author is an evangelist for organic farming. She has little understanding of scientific method. If the evidence does not favor her thesis, it is ignored" (Anon., 1951).

Ideally, the organic farm is self-sufficient in terms of needs such as fertilizers, seeds, feeds, etc. In the organic system the farm is treated as a whole entity, with an interrelationship between the soil, plants and animals in a closed recycling system. The organic farm is more generalized than the conventional farm, which tends to specialize in producing crops, hogs, eggs, milk, etc.

According to the Codex Alimentarius Commission (1999), organic agriculture is:

> a holistic production management system which promotes and enhances agroecosystem health, including biodiversity, biological cycles, and soil biological activity. It emphasizes the use of management practices in preference to the use of off-farm inputs, ... as opposed to using synthetic materials.... The primary goal is to optimize the health and productivity of interdependent communities of soil life, plants, animals and people ... the systems are based on specific and precise standards of production which aim at achieving optimal agroecosystems which are socially, ecologically and economically sustainable.

In many European countries, organic agriculture is known as ecological agriculture, reflecting this emphasis on ecosystems management. The term for organic production and products differs within the European Union (EU). In English, the term is organic; in Danish, Swedish and Spanish, it is ecological; in German, ecological or biological; and in French, Italian, Dutch and Portuguese, it is biological. The term used in Australia is organic, bio-dynamic or ecological. So, in this context, the term "organic" has a different meaning from the one we learned in chemistry class. There we learned that "organic" was used to describe a compound containing carbon. A compound not containing carbon was called "inorganic". But in relation to food, "organic" is used to describe food that has been produced in a special way: organically.

As described in the preceding paragraphs, organic farming is a production method that is intended to be sustainable and harmonious with the environment. It prohibits the use of synthetic fertilizers and pesticides, products produced by gene-modification techniques, irradiation as a preserving process, sewage sludge as fertilizer, and synthetic processing aids and feed additives. When organic herds and flocks are established, the breed or strain of animal should be selected so that the animals are adapted to their environment and resistant to certain diseases. Livestock must come from holdings that comply with the rules governing organic farming, and must be reared in accordance with those rules throughout their lives.

The main differences between organic and conventional farming that emerged from these early developments were that no chemical fertilizers or chemical pesticides can be used on organic crops, and animals raised organically have to be fed on organic or natural sources of feed. Thus organic production differs from conventional production, and in many ways is close to the agriculture of Asia.

The result is that organic food has a very strong brand image in the eyes of consumers and thus should command a higher price in the marketplace than conventionally produced food. It is, however, more expensive to produce than conventional food, therefore it is more costly to the producer and consumer.

About the same time as developments in organic food production were taking place in Europe, similar developments were evident in other countries, including the US and Australia. The publication of *Silent Spring* by Rachel Carson in 1962 was important in that it brought the issue of pesticides and the environment to the attention of the public. As a result, an increasing number of consumers began to seek out organic foods since these had been produced without the use of chemical pesticides.

Organic regulations

"Organic" is a production claim and not a food safety claim. According to the USDA there are four labeling categories for organic foods:

(1) "100 percent organic" foods contain only certified organic ingredients and use certified organic processing aids.
(2) "Organic" foods must contain at least 95 percent certified organic ingredients.

(3)   Foods "made with organic" (specified ingredients) must contain at least 70 percent organic ingredients.
(4)   Foods with less than 70 percent organic ingredients may not display the USDA "organic" seal on the package but may identify which ingredients are organic on the ingredient panel.

Other countries and organizations have derived their own standards for defining foods as organic. Organic production also requires certification and verification of the production system. This requires that the organic producer maintain records sufficient to preserve the identity of all organically managed crops and stock, and records of all inputs and all edible and non-edible organic products from the farm.

The whole organic process involves four stages:

(1)   Application of organic principles (standards and regulations).
(2)   Adherence to local organic regulations.
(3)   Certification by local organic regulators.
(4)   Verification by local certifying agencies.

Currently there is no universal standard for organic food production worldwide. As a result many countries have now established national standards. These have been derived from the standards originally developed in Europe by the Standards Committee of IFOAM (International Federation of Organic Agriculture Movements) and the guidelines for organically produced food developed within the framework of the Codex Alimentarius. Within the Codex, the Organic Guidelines include Organic Livestock Production. The pertinent regulations from several countries are listed in the References (see European Commission 1991, 1999; MAFF 2001, 2006; NOP 2000).

IFOAM Basic Standards were issued in 1998 and updated in 2005. A current review to be published in 2011 is expected to define terms such as "organic" and "sustainable". The IFOAM standard is intended as a worldwide guideline for accredited certifiers to fulfil. IFOAM works closely with certifying bodies around the world to ensure that they operate to the same standards.

The Codex Alimentarius Commission (CAC) is an international standards-setting body for food and food products that is run jointly by the UN Food and Agriculture Organization and the World Health Organization. As such, it is recognized as a standardizing body by the World Trade Organization's Agreement on the Application of Sanitary and Phytosanitary Measures. WTO member governments are required by the Agreement to base their standards on international standards, including those of the Codex Alimentarius (www.codexalimentarius.net/web/index_en.jsp). The main purpose of the Codex is to protect the health of consumers and ensure fair trade practices in the food trade, and also to promote coordination of all food standards work undertaken by international governmental and non-governmental organizations. The Codex is a world-wide guideline for state and other agencies to develop their own standards and regulations, but it does not certify products directly. The standards set out in the Codex and by IFOAM are quite general, outlining principles and criteria that have to be fulfilled. They are less detailed than the regulations developed specifically for regions such as Europe.

Although there is as yet no internationally accepted regulation on organic standards, the World Trade Organization and the global trading community are increasingly relying on the Codex, IFOAM and the International Organization of Standardization (ISO) to provide the basis for international organic production standards, as well as certification and accreditation of production systems. The ISO, which was established in 1947, is a world-wide federation of national standards for nearly 130 countries. The most important guide for organic certification is ISO Guide 65:1996, General requirements for bodies operating product certification systems, which establishes basic operating principles for certification bodies. The IFOAM Basic Standards and Criteria are registered with the ISO as international standards.

It is likely that exporting countries introducing organic legislation will target the requirements of the three large markets, i.e. the European Union, the United States (National Organic Program, NOP) and Japan (Ministry of Agriculture, Forestry and Fisheries, MAFF). Harmonization will promote world trade in organic produce. It is apparent that equivalency among the systems operating in various countries is limited. Discussions in a number of forums including FAO, IFOAM and UNCTAD (the United Nations Conference on Trade and Development) have indicated that the volume of certification requirements and regulations is considered to be a major obstacle to the continuous and rapid development of the organic sector, especially for producers in developing countries. In 2001, IFOAM, FAO and UNCTAD decided to join forces to search for solutions to this problem. Together they organized the Conference on International Harmonization and Equivalence in Organic Agriculture, in Nuremberg, Germany, February 18–19, 2002. This event was the first of its kind where the partnership between the private organic community and United Nations institutions offered a forum for public and private discussions. One of the key recommendations of the conference was that a multi-stakeholder task force, including representatives of governments, FAO, UNCTAD and IFOAM, should be established in order to elaborate practical proposals and solutions. In response, the International Task Force on Harmonization and Equivalence in Organic Agriculture (ITF) was launched on February 18, 2003, in Nuremberg, Germany. Its agreed aim is to act as an open-ended platform for dialogue between private and public institutions involved in trade and regulatory activities in the organic agriculture sector.

The Global Organic Market Access Project is an extension of the work of the International Task Force on Harmonization and Equivalence in Organic Agriculture (ITF, www.itf-organic.org). The latter documented the world situation in 2003 (UNCTAD, 2004). The group listed 37 countries with fully implemented regulations for organic agriculture and processing.

Although there is currently no universal standard for organic food production, the production process involves the same four stages in all countries (as outlined earlier in this section). The organic designation for foods is thus based on documented certification, and no test is applied to confirm that the food in question is organic. Some observers (e.g. Popoff, 2010) have viewed this aspect as a flaw in the system since cases of fraud have occurred, and it has been suggested that an objective chemical test needs to be devised to verify the authenticity of organic food. At present, organic production requires that the producer maintain records sufficient to preserve the identity of all organically managed crops and animals, all inputs of all edible and non-edible organic products produced, in order to certify that the product is indeed organic.

Organic foods are also subject to international and national standards regarding the respective food laws. For instance, in North America organic milk has to be heat-treated and fortified with vitamin D in the same way as conventional milk, and organic white flour has to be fortified in the same way as conventional flour. In time, it is expected that organic foods will be sampled in the marketplace and subjected to chemical testing as part of the foods regulations in the same way that conventional foods are monitored now.

Currently, therefore, the consumer must accept that any organic food offered for sale has been produced according to the prevailing regulations. Possible tests to prove authenticity are being researched and will be outlined in subsequent chapters. Several cases of fraud have been reported, sellers passing off regular food as the more expensive organic food. Unfortunately, there is no way of proving that food is indeed organic just by inspecting it. Authenticity has to be verified by records. As a result many consumers prefer to buy organic food directly from the grower, who is able to provide information on how the food was produced.

The difficulty in proving a product is organic has been experienced by the author of this book. Every spring my wife and I plant a few Yukon Gold potatoes in our garden because we like the texture and appearance of its flesh. Since we can never buy seed potatoes of this variety, we buy some cooking potatoes in a grocery shop and plant these. Usually we have a good crop. In 2010, however, the potatoes did not grow. When some of the potatoes were dug up to find out why, no sprouting buds or shoots could be seen. A check on the internet and a column in the local newspaper provided the answer. Table potatoes are now being sprayed with a product that prevents sprouting and makes the potatoes more attractive in the grocery store. The obvious answer was then to buy organic Yukon Golds at an organic grocery store and plant these, since no such spraying is allowed on organic produce. The potatoes we bought looked very nice and clean, each bearing an organic label and with no sign of sprouting. They cost twice as much as the regular Yukon Golds. Did they grow in our garden? Unfortunately, no. At the end of the growing season they were still not showing any shoots. When we checked back at the store the answer we got was that the store had bought them as organic produce and no, they had not been sprayed. So were the potatoes in question organic or not and had they been sprayed? These questions have to remain unanswered. We, like the average consumer, have to accept what appears on the label.

Consumer perceptions

The growth of organic farming is a response to an increased consumer demand for food that is perceived to be fresh, wholesome and flavoursome, free of hormones, antibiotics and harmful chemicals and produced in a way that is sustainable environmentally and without the use of gene-modified (GM) crops. Purchasers of organic foods believe that these products are superior to conventional foods in terms of quality and safety. What is not clear from the published data on organic foods is the extent to which these consumer perceptions are correct. A large number of studies has been conducted on this issue, particularly in relation to nutritional quality, but no clear consensus has emerged. Several authors have claimed that organic food is nutritionally superior to conventional food but, conversely, bodies such as the UK Food Standards Agency have concluded that organic foods are nutritionally similar to conventional foods. Also, the results of food monitoring programs in several countries indicate that a growing proportion of conventional foods meets or exceeds the high standards set for pesticide and chemical residues and cannot be considered as "unsafe".

The food industry, researchers and academics need to have an authoritative and up-to-date source of unbiased information on how organic production affects food quality. Some of these effects are positive, others negative. Documentation of these findings will allow the concerns of consumers to be adequately addressed, relevant marketing programs to be established and appropriate information to be disseminated. Organic producers with some technical training in nutrition or food science will also benefit from the treatment of the topic.

The need for food professionals to have access to accurate and unbiased information on organic foods was highlighted by a Michigan study by Schuldt and Schwarz (2010) which found that consumers infer that organic food is lower in calories and can be eaten more often than conventional food, even when the nutrition label conveys identical calorie content. A comment overheard by one of the authors in the checkout lane of a

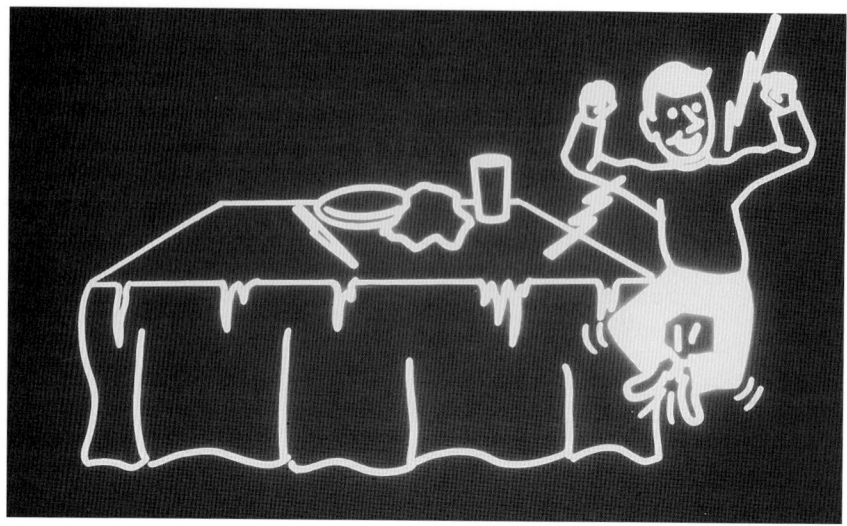

Figure 1.1   An organophile's view of organic food

natural foods store typified the confusion. "Mom, look! Organic gummy bears!" "Yes, I see. No more sweets." "Mom, but they're organic." Figure 1.1 depicts how an organophile (lover of organic food) perceives organic food.

The inference reported by Schuldt and Schwarz (2010) was more pronounced among consumers with a strong view on pro-environmentalism. Their results also indicated an "organic/natural"–"healthy" association that is capable of biasing everyday judgements about diet and exercise. Similar confusion about organic foods was found in results of an online survey of 1662 British consumers commissioned by website www.MyVoucherCodes.co.uk and reported by Halliday (2010). This website features numerous discount codes for nationwide supermarket chains online, and therefore is able to collect data on actual purchases in supermarkets. The survey found that one in four people admitted confusion and one in five believed that organic food is lower in fat. Only 16 percent said they understood the term to mean "free from synthetic chemicals". Fourteen percent said they thought it means "healthy" and 12 percent answered "expensive". Respondents were also asked questions about their purchase of diet foods, and 72 percent claimed to buy diet food regularly, three quarters of whom claimed they did so in order to lose weight. Of these, 23 percent admitted to buying 'low/reduced sugar' food as a means of weight loss, whilst 15 percent claimed to buy organic produce for the same reason. Mark Pearson, managing director of MyVoucherCodes.co.uk, is reported as commenting on the findings as follows: "The organic message has clearly been misinterpreted by a large number of the British public, many of whom seem to regard it as a diet or health food" (Pearson, 2010).

These findings raise another important issue: who should be responsible for informing consumers about the real facts relating to organic foods?

Analysis of the topic

This book addresses the topic of how organic production methods affect food quality, based on published facts. Each chapter contains the references relating to the information contained in that chapter.

In this chapter I have summarized the growth of the organic food industry and explained the motivation for its growth. This chapter also outlines how organic food is produced and certified as organic.

Chapter 2, covering consumer concerns about food, notes that consumers now have more interest in the link between food and health. Some also have concerns about the quality of food from so-called "factory farms" and question the safety of the food supply. The concerns include the possible presence of chemical and pesticide residues in food, "mad-cow disease", issues such as cloning and gene-modified (GM) foods, antibiotics, hormones, and concerns over the way plants and animals are being grown commercially as food sources and environmental sustainability.

Chapters 3 to 8 assess the documented findings related to questions concerning vegetable produce, fruit, cereal grains, meats (including fish), milk and dairy products, and eggs.

Chapter 9 asks, "Is organic food safer?" It reviews the documented evidence presented in previous chapters on the relative health aspects of organic and conventional food, based on parameters such as pesticide and chemical residues, indices of human health, and the findings of animal studies.

Chapter 10, Is organic food more nutritious and "tasty"?, discusses the documented evidence presented in previous chapters on the relative quality of organic and conventional food, including attributes such as freshness, taste and nutritional composition.

The motivation of consumers to buy organic food is explored in Chapter 11, which examines the psychology of organic food choice and presents the results of surveys. "Healthfulness" appears to be a key driver of consumer perceptions of food quality, but taste, consistency and nutritional value are also important. Of lesser importance are humane treatment of animals and environmentally sustainable production practices. Psychological issues such as the "halo effect", which are related to the choice and consumption of organic food and which may be akin to religious experiences in some people, are additional important motivating factors in the purchase of organic food.

Chapter 12 summarizes and discusses the documented findings, and makes pertinent recommendations for the various sectors of the food industry, researchers and academics.

References

Anon. (1951). Review of *The Living Soil* by E.B. Balfour. The Devin-Adair Company, New York, 1950, pp. 270. *Soil Science* **71**, 327.

Carson, R. (1962). *Silent Spring*. Houghton Mifflin, Boston.

Codex Alimentarius Commission (1999). Proposed draft guidelines for the production, processing, labelling and marketing of organic livestock and livestock products. Alinorm 99/22 A, Appendix IV. Codex Alimentarius Commission, Rome.

European Commission (1991). Council Regulation (EEC) No. 2092/91 of June 24, 1991 on organic production of agricultural products and indications referring thereto on agricultural products and foodstuffs. *Official Journal of the European Communities* **L 198**, 1–15.

European Commission (1999). Council Regulation (EC) No. 1804/1999 of July 19, 1999 supplementing Regulation (EEC) No. 2092/91 on organic production of agricultural products and indications referring thereto on agricultural products and foodstuffs to include livestock production. *Official Journal of the European Communities* **L 222**, 1–28.

European Commission (2005). Commission Regulation EC No. 1294/2005 amending Annex I to Council Regulation (EEC) No. 2092/91 on organic production of agricultural products and indications referring thereto on agricultural products and foodstuffs. *Official Journal of the European Communities* **L 205**, 16–17.

European Commission (2007). Council Regulation EC No. 834/2007 on organic production and labelling of organic and repealing regulation (EEC) No. 2092/91. *Official Journal of the European Communities* **L 189205**, 1–23.

Halliday, J. (2010). Survey shows confusion between organic food and low-fat. *Food Navigator* August 10, 2010 (www.foodnavigator.com), accessed August 11, 2010.

IFOAM (2005). IFOAM Basic Standards. International Federation of Organic Agriculture Movements, Tholey-Theley, Germany.

MAFF (2001). The Organic Standard, Japanese Organic Rules and Implementation, May 2001. Ministry of Agriculture, Forestry and Fisheries, Tokyo. http://www.maff.go.jp/soshiki/syokuhin/hinshitu/organic/eng_yuki_59.pdf, accessed January 2006.

MAFF (2006). Japanese Agricultural Standard for Organic Livestock Products, Notification No. 1608, October 27. Ministry of Agriculture, Forestry and Fisheries, Tokyo. http://www.maff.go.jp/soshiki/syokuhin/hinshitu/e_label/file/Specific JAS/Organic/JAS_OrganicLivestock.pdf, accessed October 2007.

NOP (2000). National Standards on Organic Production and Handling, 2000. United States Department of Agriculture/Agricultural Marketing Service, Washington, DC. http://www.ams.usda.gov/nop/NOP/standards.html, accessed January 2006.

Pearson, M. (2010). http://www.meattradenewsdaily.co.uk/news/060810/uk___one_in_five_think_organic_means_low_fat_.aspx, accessed May 31, 2011.

Popoff, M. (2010). *Is It Organic?* Polyphase Communication, Osoyoos, British Columbia, Canada.

Schuldt, J.P. and Schwarz, N. (2010). The "organic" path to obesity? Organic claims influence calorie judgments and exercise recommendations. *Judgment and Decision Making* **5**, 144–150.

2 Consumer Concerns About Food

After World War II there was a rapid expansion in food production, supported by advances in agricultural science. This led to an abundance of food in the developed nations.

For instance, poultry farmers learned that eggs could be produced more efficiently and cheaply if the hens were housed indoors instead of outdoors. An important feature for the farmer was that the egg production cycle could be controlled to make it even throughout the year. The bird is attuned to start producing eggs in the spring and stop laying in the winter, a hormonal response to the effects on the brain of increasing or decreasing daylight. Manipulating the light pattern in an enclosed barn allowed the farmer to simulate the natural light pattern. Consequently, egg production became a year-round system rather than a seasonal system that gave a flush of eggs in the spring and little or none in the winter.

Indoors the birds were less exposed to adverse weather, diseases and predators. As a result the farmer lost fewer birds. They also ate less because they did not have to go outside on cold days and use extra feed to keep warm.

It became common for hens to be housed in cages. This prevented losses from fighting, it being known that birds in a group develop a "pecking order" in which the bird lowest in the order suffers from bullying and is often pecked to death. Some breeds and strains of poultry are particularly aggressive.

Another benefit of cage-housing was that the birds were no longer in close contact with their droppings (urinary and fecal matter). The cages are designed to allow the droppings to fall through the cage floor into a manure pit. It is known that birds in close contact with their droppings, and especially when housed outdoors and in contact with wild birds, are more likely to be infected with salmonella. The infection can be transferred to the egg, presenting a health hazard to the human consumer. Eggs from caged birds are therefore much safer in terms of the risk of causing food poisoning. An important benefit for the consumer was that the eggs from cage-housed hens were cheaper.

That is just one example of how North American consumers have benefited from the efficiency of the modern farmer. Similar productivity improvements occurred with other commodities. As a result, the percentage of income that households spend on food is now about half what it was

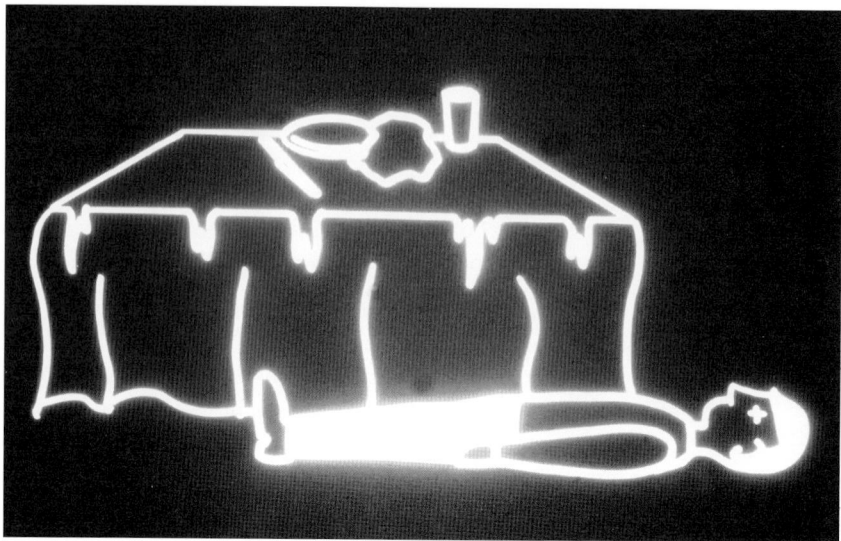

Figure 2.1   How an organophile perceives conventional food

in the early 1990s. According to 2004 statistics from the USDA Economic Research Service, American families spent on average just 9.5 percent of their disposable income on food. That means in *only 5 weeks* the average American earns enough disposable income to pay for their food supply for the entire year. The situation is the same in Canada. Consumers there spend on average 10.5 percent of their personal disposable income on food. In many parts of the world, however, food expenditure is significantly higher.

In spite of these benefits, criticisms began to be raised about the prevailing systems of food production. For example, according to their critics, these were factory farms; they inflicted cruelty on animals; they needed large amounts of chemicals and antibiotics to keep the animals healthy; food "did not taste like it used to"; crops were being sprayed repeatedly to keep down diseases, and residues of these chemicals were showing up in the food supply and threatening our health. There was also an emerging awareness of environmental conservation issues. There had to be a better way of producing food.

An important effect of these concerns was that attention turned to organic farming, which did not contain the objectionable elements of conventional farming. Figure 2.1 depicts how an organophile perceives conventional food.

The concerns

So what are these factors that are perceived to affect the quality of foods being produced by modern farming, causing consumers to turn to organic food? They fall into three main categories:

(1) *Safety*: the possible presence of harmful chemical and pesticide residues; also the possible presence of food-poisoning organisms. Included in this category is concern that the food may have been produced from plants or animals using breeding techniques such as cloning and gene modification.

(2) *Nutritional quality*: many consumers believe that food produced on a large-scale is inferior nutritionally. Many people would prefer to have their food produced locally on small farms and to have the food delivered fresh to market.

(3) *Environmental issues*: many consumers believe that organic farming is better for the environment. This issue includes concerns about the way in which animals are kept on large farms, many believing that modern production methods are cruel. Other issues raised relate to greenhouse-gas production, and the interrelationships between plants, farm animals and wildlife.

These concerns and attitudes are illustrated in the results of several studies outlined in this chapter. A more detailed assessment of the justification for these concerns, based on the relevant scientific findings, is presented in subsequent chapters dealing with specific food groups.

According to the Food & Health Survey by the International Food Information Council (2010), consumer confidence in the safety of the US food supply has remained steady for the previous 3 years, with nearly half of Americans (47 percent) rating themselves as confident in the safety of the US food supply. Those not confident fell significantly (from 24 percent in 2009 down to 18 percent) and those who are neither confident nor unconfident increased from 26 percent in 2009 to 35 percent. The survey did not provide data on attitudes towards organic foods.

Researchers in Belgium (Mondelaers *et al.*, 2009) found that health-related traits were more important than environmental traits in shaping consumer preference for organic vegetables. The presence of an organic label was important in relation to buying intensity. Undesirable traits such as pesticide residue levels triggered a stronger response than desirable traits such as environmental or health benefits.

It is clear that a main concern often voiced by consumers is the possible presence of chemical and pesticide residues in food. These include:

- herbicides used to control weeds in crops;
- insecticides and acaricides used to control insect and mite pests in crops, to protect grain, and to control external parasites on animals;
- fungicides used to control fungal diseases in plants and plant products;
- fumigants used as grain protectants, and to sterilize soil, storage sheds, barns and beehives;
- antibiotics used to control diseases in animals;

- anthelmintics used to control internal parasites in animals;
- hormonal products used as veterinary medicines or to improve growth and production in livestock.

Other sources of residues include those from the unintended exposure of plants and animals to chemicals that are no longer registered for use. Such chemicals include some organochlorine (OC) pesticides and polychlorinated biphenyl (PCB) compounds that can remain in the soil for long periods where livestock can accidentally ingest them or come into contact with them and become contaminated. In addition some chemicals in the natural environment may contaminate agricultural produce. These chemicals include certain metals and some naturally occurring mycotoxins (toxins produced by molds).

Consumer concerns and attitudes

Some of the views held by the public on chemical residues are quite startling. For instance, Williams and Hammitt (2001) surveyed over 700 purchasers of conventional and organic fresh produce in the Boston area of the United States for their perceived safety risks. This work was supported by the Harvard Center for Risk Analysis. Results showed that these consumers thought that relatively high risks were associated with the production and consumption of conventionally grown produce compared with other public health hazards. For example, those buying conventional food estimated that the median annual fatality rate due to pesticide residues on conventionally grown food was about 50 per million and those buying organic food thought that the figure was about 200 per million. As the researchers pointed out, these figures are similar in magnitude to the annual mortality risk from motor vehicle accidents in the United States! The survey also showed that over 90 percent of respondents thought that there would be a reduction in the risk of pesticide residues by substituting organically grown produce for conventionally grown produce, and nearly 50 percent thought that there would be a reduction in risk due to natural toxins and microbial pathogens by substitution.

Such views are probably the result of food scares and how these were reported in the media. For instance, beef consumption fell in the UK when the government in 1986 admitted a link between bovine spongiform encephalopathy (BSE) and human variant Creutzfeldt-Jakob disease in humans. BSE ("mad-cow disease") appears as vCJD in humans. Another effect was that the number of vegetarians in the UK doubled in the decade 1985–1995 (led by girls and young women) following identification of the epizootic (Scholten, 2006). However, as procedures were taken by regulatory authorities to ensure that no infected meat reached the public, and as reports of the disease dwindled, many consumers returned to their usual diets.

Thus, although food is now abundant, consumers in different countries have concerns about the quality of the food available, no doubt influenced mainly by recent experiences in that country. A 2007 survey in New Zealand found that 67 percent of consumers agreed with the statement, "I find product quality labels saying things like 'SPCA approved eggs' and 'free of added hormones' useful" (Table 2.1).

The survey was astute enough to include the phrase "free of added hormones" in the first question since several foods such as milk contain natural hormones.

In contrast to what was found in New Zealand, a UK survey conducted by the Food Standards Agency in 2008 found that the amount of salt in food was of greatest concern (50 percent of responses, down from 54 percent in 2006: Table 2.2). The presence of hormones in meat rated 28 percent of responses, down from 30 percent in 2006.

This suggests that, in some countries at least, the emphasis in food buying is changing to include a more positive view (nutritional aspects) as well as a negative view (risks).

Table 2.1 Public concerns about specific food issues

	Agree (percent)	Disagree (percent)	No firm opinion or unsure (percent)
I find product quality labels saying things like "SPCA* approved eggs" and "free of added hormones" useful	67	15	18
Most of the highest quality food produced in New Zealand is exported	65	10	25
I like to study the ingredient labels on food very carefully	58	20	22
I think obesity is our biggest food safety problem	55	22	23
I would like to eat organic food but it's too expensive	54	23	23
Having government regulations on food handling practices at fund-raising barbecues or food stalls is over the top	42	30	28
People that eat mainly organic food will be healthier	35	35	30
A little bit of food-related illness every now and then is good as it builds your immunity	19	59	22

* SPCA = Society for the Prevention of Cruelty to Animals.
(*Source*: Food Safety Authority (2007) Food Safety and the New Zealand Public 2007 Survey. New Zealand Food Safety Authority, Wellington, New Zealand. Reprinted with permission.)

Table 2.2 Public concerns in the UK about specific food issues

	Percent 2007	Percent 2006
The amount of salt in food	50	54
The amount of fat in food	40	46
The amount of sugar in food	39	43
The amount of saturated fat	37	44
Food poisoning	36	42
The use of additives in food	35	38
Foods aimed at children	33	39
The use of pesticides to grow food	32	34
Condition in which food animals are raised	28	32
Hormones and steroids in meat	28	30

(*Source*: Food Standards Agency (2008) Eighth Annual Consumer Attitudes to Food Survey. UK Food Standards Agency, London, UK. © Crown Copyright material reproduced with the permission of the Controller, HMSO.)

Other examples of consumer attitudes towards organic foods can be cited. Research in Ireland examined consumer perceptions of organic meat (O'Donovan and McCarthy, 2002). Purchasers of organic meat believed that it was superior to conventional meat in terms of quality, safety, labeling, production methods and value.

van Rijswijk and Frewer (2008) found regional differences in responses. German respondents defined quality primarily in terms of "freshness", "taste", "natural/organic", and as a "good product" associated with a "proper production method". French respondents linked quality with both "taste" and "appearance". Italian respondents indicated that a "good product", "taste" and "liking" are important definitions of quality. Spanish respondents defined quality as resulting in "good products", with "taste" and "without risk" (i.e. safe). With respect to safety, German and Italian respondents were most concerned about "risk" and "healthiness", whereas the Spanish respondents in particular emphasized "controls" and "guarantees" as being important determinants of safety. These associations were also expressed by French respondents, in addition to "proper handling and chain management".

One of the challenges in conducting surveys such as these is to obtain a representative sample of consumers. Interesting data were reported by Scholten (2006), who studied local and organic food consumption and risk perceptions in Seattle (Washington, USA) and Newcastle (UK). He argued that, since firefighting is risky work and since firefighters occupy a middle position on the socio-economic ladder and food is seen as a key to morale and fitness, firefighters could be viewed as suitable consumers to be consulted on risks associated with food issues. He found that firefighters in the UK and the US were aware of organic food and that firefighters' consumption of this type of food was higher in Seattle (USA), at 64 percent, than by firefighters in Newcastle (UK), where the figure was 39 percent. This survey also suggested that consumption of organic food in northern UK would increase if availability increased and the price was lower.

Scholten *et al*. (2006) surveyed another group that would not normally be regarded as typical organic consumers, namely motorcyclists. They found that a greater proportion of Seattle (US) motorcyclists (69 percent) consumed organic food than of Newcastle (UK) motorcyclists (40 percent). In comparison, 88 percent of academics (teachers and students) in Seattle and 78 percent in Newcastle consumed organic food. Scholten and co-workers also suggested that the results of their surveys dispelled the notion that women are more aware of organic food than men.

This research suggests that increasing awareness of food safety and pollution issues by a wide cross-section of consumers is an important determinant in the purchase of organic meat, causing some consumers to turn away from conventional foods. Lack of availability and the higher price appear to be key deterrents to the purchase of organic foods.

It is useful in this context to consider how consumers view food quality and safety. Van Rijswijk and Frewer (2008) found that there was considerable overlap among responses when consumers were asked to define these terms. However, quality was more frequently defined in terms of "taste", "good product", "natural/organic" and "freshness", whereas safety was primarily defined in terms of "absence of risk" or "harmfulness" whilst being positively associated with "health".

These responses will therefore be used in this book as the basis for dealing with issues of quality and safety. Quality will be taken to refer to nutritional content, and to freshness, taste and appearance which are obvious to the consumer, whereas safety will refer to perceived freedom from harmful chemical and pesticide residues, hormonal residues and "healthfulness" (ability to enhance or promote health in the consumer).

An important finding with regard to consumer attitudes to food safety is that they change as consumers become better informed of the relevant issues. This was illustrated in a study conducted in the UK (Barnes *et al*., 2009). This study was conducted in a small Scottish town to assess consumers' perception of the term "organic" and to determine what consumers would like from an organic production system, using a "citizens' jury" approach. The procedure consisted of a small group of people, selected from the general public, who met over a number of days to deliberate on a particular question. The major advantage of this approach is its ability to provide more in-depth technical information and to offer fewer time constraints on the deliberation process. In the case of dairy farming, the jury was presented with evidence of the economic, environmental and welfare impacts of both conventional and organic systems, and asked to recommend its favored system and how it would define the term "organic".

The study found that the jury's knowledge of certain aspects of farming improved over the two-day period and its members were unanimous in favoring organic agriculture when solely considering the environment, but their views towards the animal welfare effects were mixed. Conversely, when considering the economic impacts the jury supported the conventional system. A serious omission from the study was the lack of any consideration of the nutritive quality or relative safety of the foods examined. However, the approach is preferable to the questionnaire or focus-group approach, both of which impose constraints on the dissemination of information and the time allowed to deliberate on decisions.

The issue of cloning is part of the issue of food safety, but appears to be of lesser importance to the consumer than the possible presence of pesticide and chemical residues. Cloning – at least of plant material – appears to be more acceptable to the consumer than gene modification (GM). This issue will be dealt with in more detail in succeeding chapters.

Before examining the justification for the consumer attitudes outlined above, an outline of how food safety is controlled by regulatory authorities is useful in providing a background to the situation regarding residues in food.

Food regulations

Developed countries have government agencies which ensure that the food we buy is safe. The regulations set upper limits on the content of chemical and pesticide residues that are allowed in foods offered for sale at the retail level.

There are no specific food safety regulations that apply to organic foods. All foods are subject to the same regulations, including imported foods. It could be asked why any residues at all are allowed. The answer is that no food is entirely free of chemicals. The nutrients used by animals and plants are chemicals. Also, analytical methods are now so sensitive that residues even in very small amounts are likely to be detected whenever a test is conducted. The methods can detect compounds in amounts as low as parts per million (ppm, equivalent to milligrams per kilogram, mg/kg) and even parts per billion (ppb equivalent to micrograms per kilogram, μg/kg).

In most cases a maximum tolerance level is set for a chemical residue, i.e. the concentration below which no deleterious effects have been found. In other cases, however, when the residue in question is considered dangerous, no amount of residue is accepted. Controlling the food safety program requires extensive inspection, monitoring and review.

In the US the Federal Food, Drug and Cosmetic Act (FFDCA) governs the setting of chemical pesticide tolerances for food and feed products. The Environmental Protection Agency (EPA) and the Food and Drug

Administration (FDA) are responsible for administering the Act. The Food Quality Protection Act (FQPA) which was passed in 1996 set a higher standard for pesticides used on food. In addition to food, the US safety standard takes into account the total risk from all sources of exposure to chemicals, such as drinking water and residential lawn use. Under the law, EPA may only establish a tolerance if there is "a reasonable certainty" that no harm will result from all combined sources of exposure to pesticides.

Prior to the registration of a pesticide, the EPA requires testing to determine safe levels in food. Testing involves feeding high doses of each pesticide to laboratory animals to determine adverse effects (including cancer) from both acute (short-term) and chronic (long-term) exposures. From these studies a reference dose is determined, the EPA estimate of what constitutes a safe level of exposure in the diet. Based on this information and estimates of food consumption patterns for the US population, the EPA then establishes a residue level for each pesticide in foodstuffs. This safe level is called a food tolerance. Foodstuffs found to have a higher residue level than the tolerance level are in violation of the law. It is the responsibility of the FDA to test for the presence of pesticide and related residues in food and to ensure that tolerances are not exceeded.

It needs to be recognized, however, that not all decisions on food and related issues are reached on the basis of the scientific evidence alone. Decisions by committees and regulatory authorities can be swayed by interpretation of safety thresholds, trade and political considerations and lobbying. This explains, for instance, how regulations in North America, Europe and Asia on the use of implant hormones in cattle can differ even though those making the decisions have access to the same scientific data.

Justification for consumer attitudes about the safety of organic and conventional foods

Several authors have reviewed the scientific findings related to this issue. For instance, Magkos *et al.* (2006) dealt with the question of whether purchasing organic food was really buying more safety or just peace of mind. They concluded that there was an urgent need for scientific information related to health benefits and/or hazards of both organic and conventional foods, but that generalized conclusions had to be tentative because of the scarcity of adequate comparative data. Organic fruits and vegetables can be expected to contain fewer agrochemical residues than conventional products, but the significance of this difference is questionable since the determined levels of contamination in both types of food are generally well below acceptable limits. Also, some leafy, root, and tuber organic vegetables appear to have lower nitrate content compared with conventional products, but whether or not dietary nitrate indeed constitutes a threat to human health is a matter of debate. On the other hand, no differences can be identified for environmental contaminants (e.g. cadmium and other heavy metals), which are likely to be present in food from both origins. With respect to other food hazards, such as endogenous plant toxins, biological pesticides and pathogenic microorganisms, the available evidence is extremely limited and prevents generalized statements being made. Also, results for mycotoxin contamination in cereal crops are variable and inconclusive; hence, no clear picture emerges. It is difficult, therefore, to weigh the risks, but what should be made clear is that "organic" does not automatically equal "safe". Additional studies in this area of research are warranted. This issue will be examined in detail in the succeeding chapters of this book.

Range of organic foods

In response to consumer demand, organic farming is now practiced in about 150 countries throughout the world. The regions with the largest areas of organically managed agricultural land are Australia/Oceania, Europe and Latin America. There are more than 600 000 organic farms worldwide, almost half of these in Africa.

A whole range of organic foodstuffs is now available in many countries, including vegetable produce, fruit, milk, meat and eggs. An interesting case is that of organic fish. Yes, there is such a product. Much to the chagrin of some organic purists it appears that it is farmed fish that are "organic", not wild fish harvested from the sea! This is because farmed fish can be fed and reared in such a way that they meet the standards for organic designation. Since fishermen have no control over the rearing or feeding of wild fish it is impossible to grant the organic designation to their catches. This explains the labeling of fish in food stores as "wild" or "farmed".

How about some fine wine to accompany that delicious meal? Organic wines are now available, welcome, no doubt, since many conventional wines are "no better than a sort of alcoholic cola" according to Malcolm Gluck, author of *The Great Wine Swindle*. "You get artificial yeasts, enzymes, sugar, extracts, tannins, all sorts of things added." Apparently some cheap wines have oak chips added to create the impression that they have been fermented in a traditional barrel. Organic Scotch whisky is also now available, for that organic, pre-meal cocktail! One brand was good enough to win the Supreme title in the 2007 Scottish Food and Drink Excellence Awards competition. The award-winning Benromach Organic Speyside Single Malt Whisky is aged in new American casks made from oak harvested from environmentally managed forests, since regular casks cannot be certified as organic. This organic whisky is certified by the Soil Association.

References

Barnes, A.P., Vergunsts, P. and Topp, K. (2009). Assessing the consumer perception of the term "organic": a citizens' jury approach. *British Food Journal* **111**, 155–164.

Food Quality Protection Act (1996). US Food and Drug Administration, US Department of Health and Human Services, Washington, DC. http://www.epa.gov/pesticides/regulating/laws/fqpa/, accessed October 10, 2010.

Food Safety Authority (2007). Food Safety and the New Zealand Public 2007 Survey. New Zealand Food Safety Authority, Wellington, New Zealand.

Food Standards Agency (2008). Eighth Annual Consumer Attitudes to Food Survey. UK Food Standards Agency, London.

Gluck, M., (2008). *The Great Wine Swindle*. Gibson Square Books, London.

International Food Information Council (2010). *2010 Food & Health Survey: Consumer Attitudes Towards Food Safety, Nutrition, and Health*. International Food Information Council Foundation, Washington, DC.

Magkos, F., Arvaniti, F. and Zampelas, A. (2006). Organic food: buying more safety or just peace of mind? A critical review of the literature. *Critical Reviews in Food Science and Nutrition* **46**, 23–56.

Mondelaers, K., Verbeke, M. and Van Huylenbroeck, G. (2009). Importance of health and environment as quality traits in the buying decision of organic products. *British Food Journal* **111**, 1120–1139.

O'Donovan, P. and McCarthy, M. (2002). Irish consumer preference for organic meat. *British Food Journal* **104**, 353–370.

Scholten, B.A. (2006). Firefighters in the UK and the US: risk perception of local and organic foods. *Scottish Geographical Journal* **122**, 130–148.

Scholten, B.A., Holt, G. and Reed, M. (eds) (2006). Motorcyclists in the USA and the UK: risk perception of local and organic food. In *Sociological Perspectives of Organic Agriculture: From Pioneer to Policy*, pp. 107–125. Commonwealth Agricultural Bureau International (CABI), Wallingford, UK.

van Rijswijk, W. and Frewer, L.J. (2008). Consumer perceptions of food quality and safety and their relation to traceability. *British Food Journal* **110**, 1034–1046.

Williams, P.R.D. and Hammitt, J.K. (2001). Perceived risks of conventional and organic produce: pesticides, pathogens, and natural toxins. *Risk Analysis* **21**, 319–330.

3 Vegetable Produce

Including vegetables and fruit in our diets is important in helping to maintain our health. The World Health Organization (2003) has estimated that a low vegetable and fruit intake accounts for 2.7 million deaths annually. However, while the public recognizes the importance of such produce in the diet, a main concern is that conventional (regular, non-organic) produce might contain pesticide or other chemical residues that could build up in the body and cause cancer and other diseases.

Pest and disease control

Growing vegetable produce that is affordable, good-tasting, and safe to eat is challenging. It is estimated that each year insects, rodents, fungi, weeds and disease collectively damage or destroy about a third of the world's food crops. In North America alone they cause some $20 billion worth of damage (Yudelman *et al.*, 1998). Pesticides are the chief weapons that farmers use to control this damage. Some pesticides are toxic in concentrated form. As a result they have to be applied under strict controls to ensure that they are used responsibly. To ensure safety to the public, samples of foods are taken each year from supermarkets and food stores to test for residues under food monitoring programs.

A logical question often posed by the consumer is: who approves the use of pesticides and monitors their possible presence in food?

In the United States the Environmental Protection Agency (EPA) has the responsibility for regulating the development and use of pesticides. Other countries have similar agencies. Before approving or registering a pesticide for use in food production, the EPA requires adequate testing to determine its safety. The agency registers only those pesticides that meet their standards for human health and impacts on the environment and wildlife. Whenever new research findings indicate that any registered pesticide no longer meets the current standards, the agency will cancel or modify its use.

The agency also establishes a tolerance level for each approved pesticide. This is the maximum residue level of a pesticide legally permitted in or on a

Organic Production and Food Quality, First Edition. Robert Blair.
© 2012 John Wiley & Sons, Ltd. Published 2012 by John Wiley & Sons, Ltd.

food. It ensures that when the pesticide is used according to label directions, any remaining residue on food will not pose an unacceptable health risk to any consumer of the food, especially children.

EPA establishes a reference dose (RfD) for each pesticide it approves for use. An RfD is the amount of a chemical that, if ingested over a lifetime, is not expected to cause any adverse health effects in any population sub-group. The RfD includes a 10- to 10 000-fold safety factor to protect humans over a lifetime including infants, children and other special populations. Using food consumption and other data, EPA estimates how much pesti-cide residue is likely to be consumed. If the RfD is exceeded, the agency takes steps to limit the use of the pesticide.

The various tolerance levels are considered an enforcement tool and are used by the US Food and Drug Administration (FDA) in its monitoring program to ensure a safe food supply. If any pesticide residue is found to exceed its tolerance on a food, then the food is not permitted to be sold. FDA enforces pesticide tolerances for all foods, except for meat, poultry and some egg products, which are monitored by the US Department of Agri-culture (USDA). The regulations apply to both domestic and imported foods. Some states, such as California and Florida which grow much of the nation's produce and fruit, also have their own monitoring programs. Most food distributors and manufacturers monitor the farm use of pesticides to ensure that the produce meets strict quality assurance standards.

Laboratory equipment used by these agencies can usually detect residues present at 1 part per billion – the equivalent of one inch in 16 000 miles – or lower (see Appendix).

The situation in Canada is similar, with strict controls on pesticide use. Before the Pest Management Regulatory Authority (PMRA) will register a product, it must pass a rigorous safety assessment. This assessment includes a chemical evaluation, toxicological research, field testing, evalu-ation of product effectiveness and various risk versus benefit analyses. The approval process for a new pesticide or a new use of a pesticide can take several years. If a pesticide successfully completes the assessment process, it is "registered" for use and becomes subject to provincial/territorial regulations (PMRA, 2003).

Based on the assessment, dietary habits of all age groups and a lifetime of exposure to the pesticide from a variety of foods, a maximum residue limit (MRL) for a pesticide and its by-products on foods is established. This limit is many times lower than levels that would actually pose a health risk (PMRA, 2003). These minute amounts of pesticides and their by-products allowed are usually in the range of no allowable residue to several parts per million (ppm) or parts per billion (ppb). The federal government tests thousands of samples of fruits and vegetables annually to ensure that residues, if present, are within the allowable safe limits and to determine if illegal pesticides are being used. Both domestic and imported produce is tested. To date, the number of violations has been found to be low at around

2 percent. The federal government takes appropriate enforcement actions in cases of violation (PMRA, 2003).

The PMRA continually evaluates the safety of a pesticide after it is registered and used. This includes continued dietary assessments to determine total consumer exposure to pesticide residues, environmental monitoring and reviewing the latest scientific data regarding the allowable safe limits for pesticide residues (PMRA, 2003).

A development that has been introduced by the plant industry to minimize the use of pesticides on farms is integrated pest management (IPM). The intention of IPM is to work in harmony with nature by utilizing "good bugs" such as ladybugs to destroy "bad bugs", and by using other natural control methods. Pesticides are used only in limited amounts when pests reach damaging concentrations, rather than on a routine basis. Many pesticides now being developed utilize biologicals or natural substances in the environment to help destroy pests. In addition, research in plant breeding now emphasizes the development of crops that have greater natural resistance to pests and disease organisms.

The organic regulations worldwide avoid the use of most or all pesticides, chemical pest control agents and artificial fertilizers that are used in conventional agriculture and horticulture. So are organic foods safer to the consumer? Logically, any residue levels should be lower. That question is difficult to answer since there has been very little documentation of residue levels. This is an important issue because many consumers buy organic foods in the belief that they have a lower content of chemical residues. Also, there is a related question: if the residue levels are lower, does that lead to better health?

Documented findings on pesticide residues

In general, the documented findings on residue levels in foods are reassuring, in spite of consumer concerns. For instance, a 1996 report by the US National Academy of Sciences (NAS-NRC, 1996) concluded that the intake of both synthetic and naturally occurring pesticides from food is at such a low level that pesticides pose little threat to human health.

A study in New Zealand found very little evidence to show that organic foods contain lower levels of chemical residues. This was after a review of 209 relevant reports in the international scientific literature (Bourn and Prescott, 2002). Consequently, the New Zealand researchers recommended that further analytical studies needed to be conducted to clarify the issue.

On the other hand a US report published in the same year found lower levels of pesticide residues in organic vegetables and fruit than in regular produce (Baker *et al.*, 2002). This was a collaborative study involving the Organic Materials Review Institute (Eugene, Oregon), Benbrook Consulting

Table 3.1   Frequency of pesticide residues in vegetables by cultivation type, excluding the residues of banned organochlorines: USDA pesticide data program results, 1994–1999 (Baker *et al.*, 2002)

	Organic		Conventional	
	Number sampled	Percent positive	Number sampled	Percent positive
Broccoli	2	50	674	25
Carrots	18	22	1874	73
Celery	2	50	173	96
Cucumbers	10	20	723	74
Green beans	3	-	1169	59
Lettuce	3	33	860	50
Potatoes	4	25	1386	81
Spinach	9	47	1645	84
Sweet bell peppers	11	9	722	69
Sweet potatoes	6	17	1557	64
Tomatoes	10	-	1971	64
Winter squash	9	11	1205	41
All vegetables	97	23	13959	65

Services (Idaho) and the Consumers Union of the US (New York), and covered organically grown and non-organic fresh fruits and vegetables. The data analyzed were from three testing programs: the Pesticide Data Program of the US Department of Agriculture; the Marketplace Surveillance Program of the California Department of Pesticide Regulation; and private tests by the Consumers Union, an independent testing organization. The data sets were from 1989 to 1999.

The report found that organically grown foods consistently had about one-third as many residues as conventionally grown foods. In addition, conventionally grown samples were also far more likely to contain multiple pesticide residues than were organically grown samples. Comparison of specific residues on specific crops found that residue concentrations in organic samples were consistently lower than in the other category, across all three data sets. An example of the data presented in the report is shown in Table 3.1.

The data in Table 3.1 indicate that during 1994–1999 organic produce tested by USDA was lower in pesticide residues than conventional produce. This is a reassuring finding for the organic consumer. However, the amount of sampling in the organic produce was much lower than in the conventional produce, suggesting that more intensive sampling is needed and that the analysis needs to be brought up to date. Another question is how the positive levels relate to the established tolerable levels.

The data collected during 2006 in the USDA Pesticide Monitoring Program show a more reassuring picture of the quality of conventional foods produced in the USA (Table 3.2). In that year 98.2 percent of all

Table 3.2 Frequency of pesticide residues in US domestic produce, excluding the residues of banned organochlorines (USDA, 2007: http://www.ams.usda.gov/pdp)

Product	Total samples analyzed	Percent with no residues	Samples in violation	Violations over tolerance	Violations no tolerance
Corn (maize)	44	100	0	0	0
Peas (green/snow/ sugar/sweet)	20	90.0	10.0	0	2
String beans (green/snap/pole/ long)	34	73.5	0	0	0
Bean and pea sprouts	3	100	0	0	0
Other beans and peas and products	64	92.2	0	0	0
Cucumbers	25	80	0	0	0
Eggplant	7	85.7	0	0	0
Okra	5	80	20.0	0	1
Peppers, hot	0	0	0	0	0
Peppers, sweet	15	73.3	0	0	0
Squash/pumpkins	58	67.2	3.4	1	1
Tomatoes	44	79.5	0	0	0
Other fruiting vegetables	4	25	75.0	0	3
Asparagus	6	100	0	0	0
Bok choy and Chinese cabbage	2	0.0	50.0	0	1
Broccoli	9	88.9	0	0	0
Cabbage	22	77.3	0	0	0
Cauliflower	0	0	0	0	0
Celery	0	0	0	0	0
Collards	4	25	0	0	0
Endive	1	100	0	0	0
Kale	4	50	1	1	0
Lettuce, head	7	85.7	0	0	0
Lettuce, leaf	37	43.2	2.7	0	1
Mustard greens	3	33.3	0	0	0
Spinach	12	33.3	0	0	0
Other leaf and stem vegetables	18	66.7	5.6	0	1
Mushrooms and truffles	13	84.6	0	0	0
Carrots	9	55.6	22.9	0	2
Onions/leeks/ scallions/shallots	20	85	0	0	0

(*Continued*)

Table 3.2 (*Continued*)

Product	Total samples analyzed	Percent with no residues	Samples in violation	Violations over tolerance	Violations no tolerance
Potatoes	86	61.6	0	0	0
Radishes	5	0	20.0	0	1
Red beets	13	38.5	7.7	0	1
Sweet potatoes	15	86.7	0	0	0
Other root and tuber vegetables	30	70	0	0	0
Vegetables, dried or paste	20	100	0	0	0
Other vegetables/ vegetable products	13	84.6	0	0	0
Total	672	73.8	2.4	2	14

domestic foods analyzed were in compliance with established residue tolerances and formal action levels. Of the 1260 domestic samples tested, 68.8 percent had no detectable residues and 1.6 percent had violative residues. In the vegetables group 73.8 percent of the samples had no detectable residues and 2.4 percent had violative residues.

These figures present a more reassuring picture of the residue situation than in earlier years, indicating that the monitoring and testing programs are having positive effects for the consumer. However, they indicate a rate of testing that many consumers would like to see expanded. Also, no distinction was drawn between conventional and organic foods in the 2006 data.

The USDA tested 13 381 samples in 2008 and published the results in 2010. The total comprised 10 382 samples of fruit and vegetables, 186 almond, 558 honey, 552 catfish, 650 corn grain, 184 rice, 250 potable groundwater, and 619 treated and untreated drinking water samples. This included the testing of 16 fresh fruit and vegetables (asparagus, cultivated blueberries, broccoli, celery, green beans, green onions, greens (collard/kale), nectarines, peaches, potatoes, spinach, strawberries, summer squash, sweet corn, sweet potatoes, and tomatoes) and five processed commodities (apple juice, frozen cultivated blueberries, grape juice, canned kidney beans, and frozen corn).

The tests showed that, for fresh and processed fruit and vegetables, almonds, honey, catfish, and rice, approximately 76.4 percent of all samples tested were from US sources, 19.8 percent were imports, 2.7 percent were of mixed national origin, and 1.1 percent were of unknown origin.

Of the 11 960 samples analyzed, the overall percentage of total residue detections was 1.6 percent. The percentage of total residue detections for fresh fruit and vegetables ranged from 0 to 3.3 percent, with an average of

1.9 percent. The percentage of total residue detections for processed fruit and vegetables ranged from 0 to 2.2 percent, with an average of 0.6 percent. The percentage of total residue detections for almonds was 1.4 percent, for honey was 0.4 percent, for corn grain was 0.7 percent, and for rice was 0.7 percent. Low levels of environmental contaminants were detected in celery, collard and kale greens, spinach, summer squash, and catfish at concentrations well below levels that trigger regulatory actions. Residues exceeding the tolerance were detected in 0.5 percent of the 11 960 samples tested in 2008: 58 samples with one residue exceeding and two samples with two residues exceeding the established tolerance. Residues with no established tolerance were found in 3.7 percent of the samples. In most cases these residues were detected at very low levels and some residues may have resulted from spray drift or crop rotations.

Several examples illustrated the trend for a reduction in the levels of residues. The pesticide chlorothalonil was found in 29.8 percent of celery samples tested in 2008; the highest level being 0.9 ppm (Table 3.3). This level is significantly lower than in 2002 when chlorothalonil was found in 62 percent of celery samples and the highest level was 11 ppm. The amount permitted in the regulations is 15 ppm. Similarly, the pesticide dicloran was found in 37 percent of celery samples tested in 2008; the highest level being 1.5 ppm. This level is significantly lower than in 2002 when dicloran was found in 46 percent of celery samples and the highest level was 18 ppm. The amount permitted in the regulations is 15 ppm. This lowering trend was observed for several "older" pesticides for which the EPA has modified the use or has approved "safer" pesticides for use in US agriculture. Residues of organophosphates decreased from 48.7 in 2002 to 35.0 percent in 2008; residues of methamidophos decreased from 33.0 percent to 20.8 percent; and residues of older pesticides such as chlorothalonil and oxamyl decreased from 62.2 to 29.8 percent and 36.9 to 16.3 percent, respectively.

The Pesticide Data Program re-tests high consumption foods every 5 years or sooner so that the EPA has pertinent and current knowledge of residues in foods.

Residue levels have also been studied. Results of the monitoring program of pesticide residues in organic food of plant origin in Lombardy (Italy) were reported by researchers from the International Centre for Pesticides and Health Risk Prevention, Milan (Tasiopoulou *et al.*, 2007). During the period 2002–2005, a total of 3508 samples of food of plant origin were analyzed for pesticide residues. Included in the study were 266 samples of organic farming products. The data showed that the vast majority of organic farming products were in conformity with the relevant legislation and did not contain detectable pesticide residues. Only one sample showed a residue level above the maximum residue limit (MRL). It was not, however, judged to pose a concern for public health when assessed in terms of dietary risk assessment.

Table 3.3 Comparison of residues of pesticides in celery during 2002 and 2008. USDA pesticide data program results, 2010. http://www.ams.usda.gov/pdp

Pesticide	Celery 2002				Celery 2008			
	Minimum concentration detected (ppm)	Maximum concentration detected (ppm)	EPA tolerance (ppm)	Samples with detections (%)*	Minimum concentration detected (ppm)	Maximum concentration detected (ppm)	EPA tolerance (ppm)	Samples with detections (%)*
Acephate	0.003	1.8	10	48.7	0.003	0.77	10	35.0
Acetamiprid		NA			0.001	0.068	3.00	22.1
Azoxystrobin		NA			0.0008	0.74	30.0	19.6
Chlorothalonil	0.004	11	15	62.2	0.013	0.90	15	29.8
Cyromazine		NA			0.004	0.078	7.0	37.4
Dicloran	0.013	18	15	46.1	0.010	1.5	15	37.0
Flonicamid		NA			0.002	0.16	0.40	13.3
Imidacloprid		NA			0.001	0.032	6.0	27.7
Linuron	0.030	0.12	0.5	1.6	0.003	0.2	0.5	30.8
Malathion	0.003	5.5	8	25.8	0.003	0.74	8	19.3
Methamidophos	0.002	0.22	1.0	33.0	0.002	0.025	1.0	20.8
Methomyl	0.002	0.15	3	11.3	0.004	0.24	3	3.6
Methoxyfenozide		NA			0.002	0.25	25	50.4
Omethoate	0.003	0.041	2	15.1	0.004	0.11	2	17.4
Oxamyl	0.002	0.31	3	36.9	0.006	0.096	10.0	16.3
Permethrin cis	0.025	0.32	5.0	19.9	0.004	0.20	5.0	39.0
Permethrin trans	0.025	0.32	5.0	16.3	0.004	0.19	5.0	42.0
Propiconazole	0.035	0.11	5.0	12.7	0.027	0.087	5.0	21.9
Pyraclostrobin		NA			0.0018	0.24	29.0	11.9

* = Only pesticides with residue detections in at least 10 percent of samples for either year are shown.
NA = Not analyzed. No samples were analyzed for that pesticide/commodity pair.

The Italian team made comments similar to those made by other researchers on this topic.

> Organic fruits and vegetables can be expected to contain fewer agrochemical residues than conventionally grown alternatives. There is a widespread belief that organic agriculture products are safer and healthier than conventional foods. It is difficult to come to conclusions, but what should be made clear to the consumer is that "organic" does not automatically equal "safe". In the absence of adequate comparative data, additional studies in this area of research are required.

Other monitoring studies in Italy found that contamination of fruit and vegetables has been declining both nationally and internationally in recent years (Gambino *et al.*, 2007). Pesticide contamination of aubergines, peppers and tomatoes was found to be similar to the national average, but isolated examples of higher levels were found in strawberries and courgettes. Most courgettes showed almost no contamination, however.

A Dutch study compared levels of contaminants in Dutch organic and conventionally grown crops (Hoogenboom *et al.*, 2008). Both organic and conventional products contained no residues of pesticides above the legal limits, although some residues were detected in conventional lettuce.

Rossi *et al.* (2008) found no residues of pesticides in tomatoes grown organically or by conventional or IPM methods.

Further data on residues in conventional produce were reported from a Belgian study, an area of intensive agricultural production for many years (Claeys *et al.*, 2008). Estimated exposure of the Belgian consumer to pesticide residues from the consumption of fruit and vegetables was calculated by the Scientific Institute for Public Health and the Belgian Federal Agency for the Safety of the Food Chain using 2005 data. It was found that, for most pesticide residues studied, the exposure was 100 times lower than the acceptable daily intake (ADI). However, for a consumer of large amounts of produce (97.5th percentile of consumption), the intake of pesticide residues could reach 9–23 percent of the ADI. Further analysis of the data indicated that except for one pesticide (chlorpropham) the probability of exceeding the ADI from the consumption of fruit and vegetables was much lower than 0.1 percent.

Other chemical contaminants

Before outlining the facts relating to this issue it is useful as background to point out a main difference between conventional and organic production of vegetables and fruit, namely fertilizer use.

In conventional production a compound fertilizer derived from ammonia and mineral products is commonly used. It supplies the main nutrients required for plant growth – nitrogen (N), phosphorus (P) and potassium (K) – in the correct proportions for the crop being grown.

In organic production the fertilizer used is commonly animal manure that has been aged by composting, together with rotted plant debris. The composting process generates heat, which is beneficial in destroying food-poisoning organisms in the manure. Nitrogen is readily leached from manure piles during storage. Because of this and the fact that the compost is composed of a mixture of plant material, organic fertilizer is therefore likely to contain a relatively lower proportion of N than compound fertilizer. It is also more likely to be more variable in chemical composition.

Researchers in Greece (Magkos et al., 2006) conducted a critical review of the literature and found that the reported levels of chemical contamination in vegetables and fruit from both organic and conventional production were generally well below acceptable limits. Some leafy, root, and tuber organic vegetables were found to have a lower nitrate content than conventional vegetables. No differences were found for environmental contaminants (e.g. cadmium and other heavy metals), which are likely to be present in food from both organic and conventional sources. Like other reviewers, the researchers concluded that additional studies in this area of research were needed. They also concluded, "At our present state of knowledge, other factors rather than safety aspects seem to speak in favor of organic food."

Karavoltsos et al. (2008) tested the levels of cadmium and lead in organically produced foods in Greece. The mean values detected ranged from non-detectable to 53.4 ng/g and 65.0 ng/g for cadmium and lead, respectively. The highest cadmium concentrations were observed in cereals (21.7 ng/g), followed by leafy vegetables (15.4 ng/g), whereas for lead the highest concentrations were found in leafy vegetables (33.4 ng/g). Compared with certified organically grown foods the conventionally grown foods had cadmium and lead concentrations 64 and 61 percent higher, respectively. The results also showed that "uncertified" organic products contained much higher concentrations of cadmium and lead than either the certified-organic or conventionally grown foods.

Rossi et al. (2008) found that organic tomatoes had higher cadmium and lead contents but a lower copper content than tomatoes grown conventionally or by IPM methods. No residues of pesticides were detected in any of the crops.

A Dutch study compared contaminants in Dutch organic and conventional crops (Hoogenboom et al., 2008). Nitrate levels in head lettuce produced organically in the open field were much lower than those in conventional products. No differences were detected in Iceberg lettuce and head lettuce grown in a greenhouse. Organically produced carrots contained higher nitrate levels than conventional products. Both organic and conventional products contained no residues of pesticides above the legal limits, although some residues were detected in conventional lettuce. Organic products contained no elevated levels of heavy metals, such as cadmium.

Other toxic and antinutritional compounds in produce

Potato tubers contain glycoalkaloids, natural compounds produced by the potato to repel insects and other pests. These compounds can also be toxic to humans if they are present in high enough concentrations (the recommended maximum level is 200 mg/kg, fresh-weight basis).

A study in Portugal compared glycoalkaloid levels in organic and regular potatoes. No difference was observed for one potato variety but in the other variety glycoalkaloid levels were greater in conventionally grown (79.5 mg/kg) than in organically grown (44.6 mg/kg) potatoes (Abreu et al., 2007). In another study glycoalkaloid levels were reported to be higher in some varieties when grown organically. However, there were also significant year-to-year variations in these compounds (Hajšlová et al., 2005).

Several studies have demonstrated that nitrate levels are lower in organically grown leafy vegetables (Guadagnin et al., 2005; Magkos et al., 2006), because of the different fertilizer used. High intakes of nitrate from drinking water have been shown to cause methemoglobinemia in infants and it has been suggested that high nitrate intakes may form carcinogenic nitrosamines in the stomach. However, dietary nitrate may also exert a protective effect by releasing nitric oxide. Therefore, it is not certain whether high nitrate concentrations in vegetables should be considered detrimental to adults. This topic is discussed further in Chapter 9.

Hormones

Hartmann et al. (1998) tested a range of plants for hormones. Included were potatoes (steamed), wheat (whole meal), rice (parboiled, ground), soybeans (whole meal), haricot beans (dried), mushrooms, olive oil (native), safflower oil (native), and corn oil (refined). The background to the study was that some plants are known to possess hormonal activity, resulting in estrogenic effects due to isoflavones and coumestans and mainly affecting grazing animals. Steroid hormones (pregnenolone, dehydroepiandrosterone (DHEA), androstenedione, progesterone, testosterone, 17β-estradiol and estrone) were detected in the majority of foods tested, but at low concentrations. Based on findings reviewed by these authors it was concluded that the main sources of estrogens and progesterone in the human diet are milk products, but that eggs and vegetable foods contribute in the same order of magnitude to the hormone supply as meat.

No comparable studies appear to have been done on the hormones supplied by conventional and organic diets, but in view of the increased consumption of vegetable produce by humans this is a topic that merits further investigation.

Can organic produce cause food poisoning?

This seems an odd question since we commonly regard organic food as "healthy" and regular food as "unhealthy", but instances of food poisoning can and do occur from food from organic farms, just as they have from food from conventional farms. Probably many of these outbreaks go unreported. It may well be that organic farms are at a greater risk than conventional farms in this regard, because of the use of animal manure as fertilizer. Consumers need to be aware of this, especially since it is now known that washing microbe-infected produce is not effective in avoiding food poisoning. However, there is not enough documented data to allow a determination on whether organic farms are safer, less safe or equally at risk as sources of food poisoning.

Whether organic produce is more susceptible to microbial contamination is highly controversial. For instance, out of 86 commercially available organic vegetables that were tested for several enteric pathogens in one study, *Aeromonas* spp. were isolated from 34 percent (McMahon and Wilson, 2001). Other researchers have detected several *Escherichia coli* strains (not the dangerous O157:H7) and *Salmonella* spp. more frequently in organically than in conventionally grown vegetables (Bailey *et al.*, 1999; Mukherjee *et al.*, 2004, 2006), presumably because of the use of animal manure as fertilizer. To minimize the risk of contamination several countries such as the US and Canada do not allow the use of non-composted manure.

The view of the UK Food Standards Agency (FSA) is that there is currently no firm evidence to support the assertion that organic produce is more or less microbiologically safe than conventional food (UK Food Standards Agency, 2010). In addition, a review by the UK Ministry of Agriculture, Fisheries and Food (MAFF) concluded that there is insufficient information at present to state categorically whether the risk of pathogen transfer to produce on organic farms differs significantly from that associated with conventional farming practices (Nicholson *et al.*, 2000). Finally, the bulk of available evidence from comparative studies shows no significant differences in the bacterial status of organically and conventionally grown cereals (wheat, rye) and vegetable crops (carrots, spring mix, Swiss chard, salad vegetables) (Marx *et al.*, 1994; Rosenquist and Hansen, 2000; Hamilton-Miller and Shah, 2001; Ponce *et al.*, 2002, 2003; Moreira *et al.*, 2003; Phillips and Harrison, 2005).

A Dutch study compared contaminants and microorganisms in organic and conventional foods (Hoogenboom *et al.*, 2008). Organic products were analyzed for the presence of contaminants and microorganisms and for antibiotic resistance, and were compared with conventional products. None of the samples tested positive for *Salmonella* or *E. coli* O157.

An example of a food-poisoning outbreak that was linked to a farm growing organic produce is the 2006 Californian outbreak in which *E. coli* was determined to be the cause of the sickness. Sabin Russell reported on that outbreak in the *San Francisco Chronicle*, October 13, 2006. An abbreviated version of that report follows. Reviewing the circumstances of this outbreak is useful to organic farmers in helping to protect the consumer from outbreaks of food poisoning. The information is also useful to consumers in teaching them what food sources to avoid and what questions to ask when buying produce.

> Samples of cattle manure on pastures surrounding a spinach field have tested positive for the same strain of E. coli bacteria that killed at least three people and sickened nearly 200 others – the first direct evidence linking a Salinas Valley farm to the outbreak that has spanned 26 states and one Canadian province.
>
> The pasture is part of an unidentified beef cattle ranch that also leases its fields to spinach growers. Fences on the cattle operation had been penetrated by wild pigs, and disease detectives are trying to determine whether feral swine might have played a role in spreading the bacteria from pasture to spinach field.
>
> Tests at the California Department of Health Services laboratory in Richmond found that E. coli O157:H7 bacteria detected in three cow pies produced the same genetic fingerprints as the strain found in human victims and in the bags of spinach they had purchased.
>
> "This is a significant finding," said Kevin Reilly, deputy director of prevention services for the California Department of Health Services, during a telephone press conference Thursday afternoon.
>
> Out of nine outbreaks of E. coli food poisoning traced to spinach or lettuce from the Salinas Valley since 1995, this is the first in which investigators have been able to link the bacterial strain that caused the illness to a farm where greens were grown. Reilly stressed that the test results do not prove that the manure was responsible for the outbreak and that investigators are continuing to look at other potential sources ... All victims had eaten spinach traced to Natural Selection Foods which had bagged the fresh produce at its San Juan Bautista plant. Using plant records, the investigators then focused on

nine farms that had supplied spinach to a batch processed by Natural Selection for the Dole Baby Spinach label. Since then, they have taken more than 650 specimens from soil, water and manure on the nine farms.

Health officials exploring other possible routes of transmission have not ruled out contamination of irrigation water, improper farmworker sanitation, or bacteria spread through tainted fertilizer or dirty farm implements.

Reilly said that the farm where the suspect manure was found did not fully adhere to the voluntary guidelines used by growers to keep fresh leafy greens safe from contamination. For example, although no evidence indicates that the beef cattle had strayed onto the spinach fields, Reilly said there is concern about the proximity of cattle to spinach fields and that fences on the farm had failed to keep wildlife from trudging over pasture and fields.

"On this ranch there is a very large population of wild boar," said Reilly. There was evidence that the pigs had torn through fencing or burrowed under it, he said. "We don't know if that is the source [of contamination of the fields], but it is a potential source."

The E.-coli-positive cow pies were taken some distance away, between half a mile and a mile from the field that produced spinach suspected of sickening consumers.

Reilly said it was not uncommon for ranches in San Benito and Monterey counties to have spinach fields adjacent to cattle operations. In this case, "the field is, frankly, surrounded by pasture," Reilly said.

Federal regulators are concerned about the practice of raising cattle near fields that grow salad greens. "The relationship of farm animals to produce is certainly something to take under consideration," said Dr. Robert Brackett, branch director of the Center for Food Safety and Applied Nutrition at the Food and Drug Administration. As the industry scrambles to develop stricter guidelines – and as government agencies face political pressure to make the rules mandatory – standards such as minimum distance, upslope and downslope between pasture and fields are likely to be established.

Sabin Russell, Chronicle Medical Writer, *San Francisco Chronicle*, October 13, 2006

There is an interesting aftermath to that episode, as reported by George Raine in the *San Francisco Chronicle* on August 22, 2008, namely that the FDA would now permit irradiation of spinach and lettuce:

Nearly two years after E. coli bacteria traced to California-grown spinach killed three people and sickened 205, the federal government says it will allow producers of fresh iceberg lettuce and spinach to use irradiation to control food-borne pathogens and extend shelf life.

The Food and Drug Administration is amending the food-additive regulations to provide what it calls the safe use of ionizing radiation for just the two leafy greens. The FDA also has received petitions seeking permission to use irradiation for other lettuces and many other foods.

The government is allowing the practice in the wake of the major E. coli outbreak in 2006 and numerous other problems with food safety and recalls. But this won't be first time such a technique has been used on food. Consumers have eaten irradiated meat for years.

Despite some consumer concern, the FDA says irradiation is safe.

"The agency has determined that this action is of a type that does not individually or cumulatively have a significant effect on the human environment," reads the FDA's final rule, released Thursday and effective today.

As expected, criticism of the FDA was swift.

Food & Water Watch, a nonprofit consumer rights group that challenges what it calls corporate control and abuse of food and water resources, said that very little testing has been conducted on the safety and wholesomeness of irradiated vegetables. The group also said the action was off target.

"It is unbelievable that the FDA's first action on this issue is to turn to irradiation rather than focus on how to prevent contamination of these crops," said Wenonah Hauter, executive director of Food & Water Watch. "Instead of beefing up its capacity to inspect food facilities or test food for contamination, all the FDA has to offer consumers is an impractical, ineffective and very expensive gimmick like irradiation."

On the industry side, there is little demand for irradiation from California growers and shippers of spinach and iceberg lettuce.

"I think that from a growers' perspective, we have to consider anything that helps us provide safety for consumers, but whether this takes off depends on consumers," said Cathy Enright, vice president for government affairs for Western Growers, which represents growers, packers and shippers of nearly half of the nation's fresh fruits, vegetables and nuts.

"In any marketing decision, we have to look at cost in adapting the technology and consumer acceptance," which will take time to develop, she said.

The petition for the voluntary use of ionizing radiation was filed in 2000 by the Grocery Manufacturers Association. At the time, said Robert Brackett, the group's chief scientist, the grocers wanted permission to use irradiation in the preparation of many foods. However, they amended the petition and asked the government to focus on iceberg lettuce and spinach after the 2006 E. coli outbreak.

Spinach virtually vanished from grocery stores as demand plummeted.

"That was a big motivation for us," said Brackett, in Washington, D.C.

California producers of leafy greens, in the aftermath of the case of the contaminated spinach, formed a voluntary group called the Leafy Green Marketing Agreement, which developed a food safety protocol for its members – nearly all of the major leafy green producers in California. The approved business practices range from accommodating fieldworker sanitation to preventing animal contamination of leafy green vegetables.

The grocers' association's Brackett said, "It's more of a safety net. No matter how good a job you do with preventative steps – good practices, proper sanitation – there is still a small chance for contamination. This takes care of those small chances."

The California spinach was contaminated by feral swine, an investigation later found.

William Marler, a Seattle lawyer representing victims of food-borne illness, is handling lawsuits for 103 families affected by the outbreak.

Marler said the ionizing radiation tool "gives potential consumers more choice." He said most of the E. coli problems in recent years have been with mass-produced, bagged product, "and those products are ripe for using some kill step like irradiation to make it safer."

San Francisco Chronicle, October 13, 2006

As stated earlier, the main reason for suggestions of a higher potential risk of food poisoning from organic produce is the use of manure as fertilizer. However, a study of spring mix (mesclun) from California demonstrated average and similar populations of bacteria and molds on organic and conventional greens (Phillips and Harrison, 2005). Of 13 samples found to contain *E. coli*, nine were from conventional fields. The prevalence of *E. coli* on certified organic produce was not statistically different from that on conventional produce.

Hoogenboom *et al.* (2008) reported *Salmonella* in 30 percent of pig manure samples from 30 organic farms, similar to the incidence on conventional farms. On farms that switched to organic production more than 6 years previously, no *Salmonella* was detected, with the exception of one barn with young pigs recently purchased from another farm. No *Salmonella* was detected in manure at the nine farms with organic broilers, but it was detected at one out of ten farms with laying hens. This is comparable with conventional farms where the incidence for *Salmonella* is probably around 10 percent. *Campylobacter* was detected in manure at all organic broiler farms, being much higher than at conventional farms. The conclusion of the study was that the organic products investigated were equivalent to conventional products in terms of microbial food safety.

Another possible risk factor for organic farms is increased contact with wild animals, rodents, insects and birds, which are potential sources of contamination with fecal pathogens. Dutch research, for instance, demonstrated the presence of *Salmonella* and *Campylobacter* in rodents on organic farms (Meerburg *et al.*, 2006). Infections were found in house mice (8 of 83 *Campylobacter* positive and 1 of 83 *Salmonella* sp. strain Livingstone positive) and brown rats (1 of 8 *Campylobacter* positive) but not in other species.

However, all types of farm have to be vigilant to minimize food contamination from these sources. Effective on-farm rodent management is necessary on all farms.

An assessment of the relationship between risks due to microbes and organic food was made by the Food Safety Network at the University of Guelph, Ontario, Canada (Blaine, 2007):

> No comprehensive data exists linking specific on-farm production practices to human illness, although case studies and anecdotes abound.
>
> An episode of the popular US television show 20/20, in 2000, sparked a fierce debate over the microbial safety of organically grown fresh fruits and vegetables. Are organic foods safer than conventional foods? On the show, correspondent John Stossel concluded that organic produce was no safer than conventional produce and might in fact be more dangerous because of the heavy use of manure in organic farming (Ruterberg and Barringer, 2000). Such statements have been supported by several prominent food scientists (Tauxe, 1997; Forrer *et al.*, 2000) while the organic industry has argued that their strict standards on manure usage reduces such risks (DiMatteo, 1997). The organic industry has refrained from making direct claims of improved microbial food safety. Katherine DiMatteo, president of the Organic Trade Association, has stated publicly that "Organic is not a food safety claim" (Juday, 2000). But such claims are unnecessary for an industry which has been described as the fasted [*sic*] growing sector within the entire Canadian agri-food system. Growing consumer demand for organic foods has been estimated at 25 percent per year, mainly as a result of growing consumer concern over the human health and environmental effects of genetic engineering and chemical pesticides as well as perceptions that organic foods are healthier, tastier and safer.
>
> But what little scientific evidence there is indicates that the risks from conventionally and organically produced food are essentially the same. Efforts are needed to reduce levels of risk at the farm level for both systems.

Food Safety Network at the University of Guelph, Ontario, Canada
    (Blaine, 2007, reprinted with the kind permission of the author)

Although several food-borne disease outbreaks have been associated with organic food (Cieslak *et al.*, 1993; Tschape *et al.*, 1995), it is clear that the organic farming methods are not always at fault. Contamination of produce can occur in the field or orchard, during harvesting, post-harvest handling, processing, shipping or marketing, or in the home. For example, in a case of contamination of organically grown spinach with salmonella, the spinach tested negative for the presence of pathogens when it was harvested and on arrival at the processing plant, but positive after packaging (Magkos *et al.*, 2006). Another incident was due to cross-contamination in the laboratory. *E. coli* O157 was isolated from one sample of organic mushrooms, which were immediately withdrawn from sale by the retailer. Further microbiological investigations, however, proved that the mushroom sample was not the source of the pathogen, but became contaminated in the laboratory by a strain of *E. coli* O157 used for quality control testing (Magkos *et al.*, 2006).

A variety of unlikely foods may become infected with food-poisoning organisms. For instance, black pepper was recently identified as the likely cause of a *Salmonella* outbreak traced to salami (not organic) (Scott-Thomas, 2010). Rhode Island food safety authorities were reported as having found *Salmonella* in black pepper from Vietnam. According to the report, the company in question recalled 1.24 m pounds of salami on January 23 and a further 17 235 pounds on January 31 (2010) after a *Salmonella* outbreak occurred in more than 200 people across 42 states and the District of Columbia.

Nutrient concentrations

A strong perception held by many consumers is that organic produce is more nutritious than produce grown conventionally. However, as with fruit, data on this issue are very limited (Doyle, 2006). Also it is not clear from some reports whether the composition was studied on a comparable moisture basis.

Several reviews of the available data have been published. For instance, Hoefkens *et al.* (2009) compared the nutrient and contaminant contents of organic and conventional vegetables and potatoes, based on reports in the international scientific literature. The study included carrots, tomatoes, lettuce, spinach and potatoes. They concluded that, from a nutritional and toxicological point of view, organic vegetables and potatoes were in general not significantly different from conventional vegetables and potatoes. For some nutrients and contaminants, organic vegetables and potatoes scored significantly higher but for others they scored significantly lower. As concluded by other researchers, they found that more data from controlled paired studies are needed. Also they concluded that the majority of studies to date are of moderate or poor quality.

Recent research findings help to clarify the issue. Kristensen *et al.* (2008) measured the content of trace minerals in carrots, kale, mature peas, apples and potatoes, grown either organically (without pesticides) or conventionally (with the use of pesticides). In addition the researchers investigated whether there were differences in the retention of nutrients from organically grown produce fed to rats. The study used fruits and vegetables grown in three different ways:

(1) Low input of nutrients from animal manure and no pesticides (organic).
(2) Low input of nutrients from animal manure plus levels of pesticides allowed by regulation (organic plus).
(3) High input of nutrients from mineral fertilizers and pesticides, up to legally permitted levels (conventional).

The crops were grown in similar soil, on adjacent fields, and at the same time so that they experienced the same weather conditions. The organic crops were grown on established organic soil, and all the crops were harvested and treated in the same way. For the compositional study the produce was freeze-dried immediately after harvest. In the retention study the produce was fed to rats over a 2-year period, and dietary intake and excretion (in urine and feces) of major and trace minerals were assessed. The minerals studied were calcium, phosphorus, magnesium, sodium, potassium, iron, zinc, copper, molybdenum, cobalt, cadmium, and vanadium. Some differences were found in the content of these minerals present in the produce, on a dry-matter basis, but they were not of nutritional significance (Table 3.4). Also, no consistent differences were found in the retention of elements by rats fed diets based on the produce grown under the different cultivation systems (Table 3.5).

Kristensen *et al.* fed the produce to animals and measured the biological responses, an approach with a great deal of merit in helping to establish possible differences in the nutritive value of organic and conventional foods. There was some evidence that the rats retained different levels of the nutrients in the different types of produce (Table 3.5). Rats on the conventionally based diet showed a higher retention of magnesium than rats on the organically based diet, whereas rats on the organically based diet showed a higher retention of iron, cobalt and vanadium than rats on the conventionally based diet.

Rossi *et al.* (2008) found that organic tomatoes contained more salicylic acid but less vitamin C and lycopene than tomatoes grown using conventional and IPM methods. Also, organic tomatoes had higher cadmium and lead contents but a lower copper content.

A study in Poland compared the yield and nutritional composition of red peppers grown organically or conventionally over a 3-year period

Page 44 — Organic Production and Food Quality

Table 3.4 Contents of major and trace elements in dry matter of carrots, kale, peas, potatoes and apples grown organically and conventionally and harvested in 2001 and 2002 (means of triplicate analyses of each sample, from Kristensen *et al.*, 2008). © John Wiley & Sons

Element (g/kg)		Carrot O	Carrot C	Kale O	Kale C	Pea O	Pea C	Potato O	Potato C	Apple O	Apple C
Ca	2001	2.9	3.0	15.2	14.8	1.9	1.7	0.3	0.3	0.3	0.3
	2002	2.7	2.9	20.7	17.6	1.9	1.9	0.2	0.2	0.3	0.3
Mg	2001	0.7	0.7	1.2	1.0	1.4	1.3	0.8	0.6	0.3	0.3
	2002	1.0	0.9	1.1	1.0	1.6	1.5	0.9	0.9	0.3	0.3
P	2001	2.2	2.1	4.0	3.7	4.7	4.6	2.2	1.9	0.5	0.5
	2002	2.2	2.6	3.5	3.5	5.4	5.8	2.0	2.1	0.6	0.6
K	2001	18.9	22.2	14.3	13.8	10.3	10.2	10.4	8.9	5.0	5.6
	2002	17.0	28.3	17.0	19.2	12.9	13.5	14.9	15.3	6.5	7.2
Element (mg/kg)											
Na	2001	2142	1540	744	610	111	102	160	109	78.9	67.3
	2002	3220	1638	745	558	140	132	497	450	87.9	50.6
Cu	2001	3.3	3.2	3.1	2.7	7.5	6.6	3.9	4.1	1.9	2.0
	2002	3.7	4.2	2.8	2.7	7.0	6.8	4.5	4.7	1.7	2.0
Fe	2001	29.2	30.3	104	90.7	54.5	57.4	19.1	17.1	14.6	14.7
	2002	39.7	42.2	64.0	70.0	58.0	62.3	13.0	14.5	7.5	7.5
Mn	2001	6.7a	8.2b*	18.5	21.5	14.9	12.2	6.3	5.6	2.0	2.1
	2002	8.2	10.0	29.2	38.7	17.3	17.0	6.2	6.6	2.3	3.1
Zn	2001	9.2	10.5	12.5	11.9	29.7	31.0	11.0	10.7	1.9	1.5
	2002	12.2	10.7	10.8	10.3	36.6	31.2	9.0	11.0	4.0	1.1
Element (μg/kg)											
Mo	2001	144	66.4	2115a	1190b*	3177a	1940b*	282a	147b*	24.9	22.0
	2002	92.9	98.1	1670	465	2913	1870	524	363	43.8	56.0
Cd	2001	289	339	132	99.8	18.1	16.9	44.6a	67.7b*	4.6	5.2
	2002	333	509	264	268	30.9	23.2	36.5	58.6	–	–
Co	2001	10.2	8.9	35.1	35.8	142	74.8	–	–	–	–
	2002	17.9	23.9	89.4	117	113	124	17.7	–	–	–
V	2001	40.5	42.3	125	112	13.2	16.7	–	–	–	–
	2002	58.6	65.2	40.7	68.3	–	–	–	–	–	–

* O = Organic, C = conventional. For each crop, different letters within a row indicate a difference between cultivation methods independent of year (P < 0.05).

(Szafirowska and Elkner, 2009). In the first 2 years conventional production resulted in a higher total and marketable fruit yield (Table 3.6). In the third year a slightly higher yield was obtained with organic production. The number of peppers per plant was consistently higher with conventional production, although the large differences reported did not achieve statistical significance.

Table 3.5 Retention of minerals by rats fed diets based on ingredients grown organically and conventionally (from Kristensen *et al.*, 2008). © John Wiley & Sons

Element	Daily intake		Retention (% of intake)	
	Organic	Conventional	Organic	Conventional
Ca (mg)	62.7	65.1	18.1	18.5
Mg (mg)	9.2	9.6	17.9a	23.7b
P (mg)	30.0	32.3	12.2	12.3
K (mg)	84.5	91.5	20.9	24.1
Na (mg)	5.4	5.7	30.9	30.4
Cu (μg)	53.0	54.3	4.5	4.5
Fe (μg)	799	783	16.0a	11.5b
Mn (μg)	127	121	8.3	9.6
Zn (μg)	198	212	13.1	13.4
Mo (ng)	14.7	10.1	18.3	12.4
Cd (ng)	529	641	0.8	0.7
Co (ng)	912	650	13.2a	0.1b
V (ng)	465	368	39.1a	27.2b

a,b Values in a row with different superscripts are significantly different at $P < 0.05$.

The overall data (Table 3.7) showed that peppers from organic production were heavier and with a lower wastage than peppers produced conventionally. Color intensity was higher in the peppers produced conventionally. However, none of the differences reported achieved statistical significance.

Effects on the nutritional and chemical composition are shown in Table 3.8 (fresh-weight basis). No information was presented on the dry-matter content of the peppers. The data showed that in most years the

Table 3.6 Effect of organic vs. conventional production on the yield (kg/10 m^2) of red peppers

Cultivation method	Total yield			Diseased fruits			Marketable yield			Number of fruits per plant		
	2006	2007	2008	2006	2007	2008	2006	2007	2008	2006	2007	2008
Organic	20.8	24.6	22.6	3.7	1.0	3.8	17.1	23.6	18.8	1.5	3.1	3.3
Conventional	32.5	29.3	21.8	3.0	0.4	1.4	29.5	29.9	20.4	3.5	4.9	4.8
Significance of difference	P < 0.05	NS	P < 0.05	NS	NS	NS	P < 0.05	NS	P < 0.05	NS	NS	NS

NS = Not significant at $P < 0.05$.
(*Source:* Szafirowska, A. and Elkner, K. (2009). The comparison of yielding and nutritive value of organic and conventional pepper fruits. *Vegetable Crops Research Bulletin* **71**, 111–121. Reprinted with the kind permission of the authors.)

Table 3.7 Effects of organic vs. conventional production on the physical quality of red peppers. Data averaged over 3 years

Cultivation method	Pepper wt (g)	Waste (g)	Color index	
			Redness	Yellowness
Organic	146.3	14.3	25.3	8.8
Conventional	130.3	15.0	24.8	9.4
Significance of difference	NS	NS	NS	NS

NS = Not significant at $P < 0.05$.
(*Source:* Szafirowska, A. and Elkner, K. (2009). The comparison of yielding and nutritive value of organic and conventional pepper fruits. *Vegetable Crops Research Bulletin* **71**, 111–121. Reprinted with the kind permission of the authors.)

peppers produced organically had a higher content of vitamin C, β-carotene, soluble phenols and total flavonoids.

The results confirmed similar findings in a previous report from the same research group (Szafirowska and Elkner, 2008). Their results suggest that peppers respond to the different fertilizing and plant-protection methods of organic production by yielding a lower crop of larger peppers with a slightly lower color intensity but containing a higher level of vitamin C and other compounds of nutritional interest.

Organic production avoids the use of pest-control chemicals, leading to speculation that plants grown organically respond by increasing their production of compounds such as phenolics to aid in their protection from insects and other pests. Phytochemicals, such as phenolic compounds and vitamin C which have antioxidant properties, may be partly responsible for the health benefits associated with the consumption of fruit and vegetables. As a result several researchers have investigated the concentration of these compounds in organic produce. For instance, Asami

Table 3.8 Effect of organic vs. conventional production on the nutritional and chemical composition of red peppers

Cultivation method	Vitamin C (mg/100 g)			β-carotene (mg/100 g)			Soluble phenols (mg/100 g)			Total flavonoids (mg/100 g)		
	2006	2007	2008	2006	2007	2008	2006	2007	2008	2006	2007	2008
Organic	240	193	176	1.1	1.05	1.05	145	154	150	18.5	17.0	16.0
Conventional	220	192	167	0.84	1.1	1.01	135	142	149	17.5	16.2	15.2
Significance of difference	P < 0.05	NS	P < 0.05	P < 0.05	NS	NS	P < 0.05	P < 0.05	NS	P < 0.05	P < 0.05	NS

NS = Not significant at $P < 0.05$.
(*Source:* Szafirowska, A. and Elkner, K. (2009). The comparison of yielding and nutritive value of organic and conventional pepper fruits. *Vegetable Crops Research Bulletin* **71**, 111–121. Reprinted with the kind permission of the authors.)

et al. (2003) reported increased concentrations of vitamin C and total phenolics in frozen organic corn compared with frozen conventional corn (3.2 vs. 2.1 and 40 vs. 25 mg/100 g fresh-weight basis). However, organic production does not appear to result in a similar effect in all plants. Young *et al.* (2005) measured the content of nine major phytochemical compounds in lettuce, collards and pak choi and found that organic production resulted in a higher concentration of phytochemicals in pak choi but not in lettuce or collards.

Higher concentrations of antioxidants have been reported in tomatoes (e.g. Chassy *et al.*, 2006), a food item that is a major source of phenolic compounds in the American diet. Caris-Veyrat *et al.* (2004) conducted a study to measure the content of antioxidants in three cultivars of tomatoes grown conventionally and organically. These French researchers extended their study to include a test of the effects of tomato purée on the antioxidant plasma status of humans. They found that organic tomatoes had a higher dry-matter content (by about 10 percent) than conventional tomatoes, therefore the compositional details were expressed in terms of both fresh-weight and dry-weight basis. When the results were expressed on a fresh-weight basis, organic tomatoes had higher contents of vitamin C, carotenoids and polyphenol contents (except for chlorogenic acid) than conventional tomatoes (Table 3.9). When the results were expressed on a dry-matter basis, no significant difference was found for lycopene and naringenin. No difference in carotenoid content was found in tomato purées made from the organic and conventional tomatoes, whereas the concentrations of vitamin C and polyphenols remained higher in purées made from organic tomatoes.

In the supplementation study 20 young female volunteers were randomly divided into two groups and asked to drink during their lunch or dinner 100 g/day of tomato purée. The researchers were careful to ensure that no dietary supplements of any kind were being taken during the month

Table 3.9 Concentration of micro-constituents in purée from tomatoes grown conventionally or organically, mg/100 g, fresh-weight basis

Micro-constituent	Conventional	Organic	Significance of difference[1]
Lycopene	15.57 ± 2.19	13.54 ± 0.60	NS
β-carotene	3.56 ± 1.02	1.71 ± 0.32	NS
Vitamin C	22.53 ± 1.07	39.95 ± 0.44	$P < 0.0001$
Chlorogenic acid	7.2 ± 0.2	10.6 ± 0.3	$P < 0.001$
Rutin	2.30 ± 0.04	9.65 ± 0.28	$P < 0.0005$
Naringenin	4.83 ± 0.14	6.18 ± 0.18	$P < 0.005$

[1] NS = Nonsignificant at $P < 0.005$.
(*Source*: Caris-Veyrat *et al.*, 2004. © American Chemical Society.)

Table 3.10 Effect of organic production on the composition of potatoes

	Organic	Conventional	Statistical significance
Dry matter (%)	20.79	20.80	NS
Total protein (g/100 g fresh wt)	1.65	1.52	*
Soluble protein (g/100 g fresh wt)	0.68	0.65	NS
Total amino acids (mg/100 g fresh wt)	906.2	987.7	**
Total carbohydrates (g/100 g fresh wt)	16.17	16.50	NS
Starch (g/100 g fresh wt)	14.63	14.78	NS
Starch/carbohydrates (%)	90.26	89.27	NS
Sucrose (g/100 g fresh wt)	0.50	0.56	NS
Fructose (g/100 g fresh wt)	0.40	0.50	**
Glucose (g/100 g fresh wt)	0.65	0.66	NS
Reducing sugars (g/100 g fresh wt)	1.05	1.16	NS

* Significant at $P < 0.05$, ** significant at $P < 0.01$, NS = not significant at $P < 0.05$.
(*Source*: Maggio, A., Carillo, P., Bulmetti, G.S., *et al.* (2008). Potato yield and metabolic profiling under conventional and organic farming. *European Journal of Agronomy* 28, 343--350. © 2007 Elsevier BV. All rights reserved.)

preceding the start of the study or during the study period. The subjects did not know whether they drank the organically produced tomato purée or the conventional purée. After 3 weeks of supplementation with purée, their plasma was tested and no significant difference was found in the concentration of β-carotene, lycopene or vitamin C. All subjects showed a similar increase in plasma concentration of these components. Results obtained after a 3-week depletion period also showed no significant difference between the two groups. This study appears to be the first in which organic and conventional foods were compared in terms of their ability to enhance the nutritional status or health of consumers. Although the French study was limited in its scope, the design was very commendable and should be adopted by other researchers.

Most of the research on possible differences in nutritional content of produce grown organically and conventionally has been conducted on potatoes. Typically, organic production results in a reduction in yield which is attributed to the difference in fertilizer use. Maggio *et al.* (2008) reported a reduction in marketable yield of 25 percent, and a reduction in the percentage of large tubers (from 16.6 to 12 percent). No significant differences in chemical composition (total carbohydrates, starch, sucrose, glucose, reducing sugars, dry matter) were recorded (Table 3.10) apart from an increased content of fructose in the conventional potatoes (0.5 vs. 0.4 g/100 g, fresh-weight basis). Some differences in protein content were recorded. The concentration of total protein was significantly higher in the organic potatoes (1.65 vs. 1.52 g/100 g, fresh-weight basis) and the concentration of total amino acids was significantly higher in the conventional potatoes

(989.7 vs. 906.2 mg/100 g, fresh-weight basis). Nitrate concentrations were not recorded.

Hajšlová *et al.* (2005) conducted a 4-year study on the content of micronutrients, metals and secondary metabolites, and on organoleptic properties of potatoes grown organically and conventionally. The potatoes were from eight potato varieties. The parameters included nitrate, trace elements (arsenic, cadmium, cobalt, copper, iron, mercury, manganese, nickel, lead, selenium, zinc), vitamin C and potato glycoalkaloids, as well as chlorogenic acid, polyphenol oxidase and the rate of tuber enzymic browning. The results indicated that, in general, a lower nitrate content and a higher vitamin C and chlorogenic acid content were the parameters most consistently different in organically produced potatoes. Elevated concentrations of glycoalkaloids were also observed in some potato varieties grown organically. However, the differences were not consistent, the researchers finding that variation from year to year, variety of potato used and geographical variation had equally or more important effects on potato quality.

Bagnaresi *et al.* (2007) reported that the main differences likely to occur between organically and conventionally grown potatoes are dry-matter, total protein and ascorbic acid contents, but these appear to be linked more closely to genetic and environmental factors than to agricultural techniques.

Haase *et al.* (2007) studied the suitability of organic potatoes for industrial processing, in particular selected quality parameters at harvest and after storage. Three factorial field experiments were conducted in two consecutive years to examine the effects of preceding crop, pre-sprouting, nitrogen and potassium fertilization, and cultivar on quality attributes of organically grown potatoes destined for processing into French fries or chips. Tuber dry-matter (DM) concentration, glucose and fructose concentrations, as well as the color of crisps and the quality score of French fries, were assessed at harvest and after a 4-month storage period. It was found that tubers from organic production had a sufficiently high tuber DM concentrations for processing into French fries without impairing the texture of the fries when concentrations exceeded 23 percent. Dry-matter concentrations of tubers for chips (cv. Marlen) fell short of the required minimum of 22 percent when a combined nitrogen and potassium fertilizer was applied. The medium-early cv. Agria and medium-late cv. Marena proved to be best suited for processing into French fries under conditions of organic farming. A consistently high chip quality was achieved by the medium-early cv. Marlen, with L-values (measures of the rate of enzymic browning) of 70.8 and 66.7 at harvest and after storage, respectively. Overall, results show that the quality variables were mainly affected by cultivar, season, storage and their interaction.

One piece of information that would help to explain a possibly higher level of nutrients in organic crops is the total harvested yield, which is not measured in most studies. To take an analogy: if you want to win a prize at a flower show the strategy is to reduce the number of blooms on a plant to just a few. The plant then puts all of its energy into producing just a few large, showy blooms. Organic production tends to yield smaller harvests because of a different fertilizer regime. If the crop does have a higher nutrient content, can this be explained by the plants producing a similar amount of nutrients as conventional plants, but putting them into fewer fruits, tubers or grains? Measurement of the total harvest would allow this possible explanation to be explored, as was done in the Polish study on peppers discussed earlier (see Table 3.6 and Szafirowska and Elkner, 2009). The explanation is not, of course, of paramount importance to the consumer but is of importance to scientists and the organic industry. If organic produce can be shown conclusively to have a higher nutrient content than conventional produce, the enhancement has to have an explanation.

One research team that did report total yield as well as nutritional content was Warman and Havard (1997), at the Nova Scotia Agricultural College in Canada. The test involved two treatments, organic and conventional, for carrots (*Daucus carota* L. cv. Cellobunch) and cabbages (*Brassica oleracea* L. var *capitata* cv. Lennox), in each of 3 years. The addition of pesticides, lime and NPK (nitrogen, phosphorus, potash) fertilizer to the conventional plots followed soil test and provincial recommendations. Lime, composted manure and insect control applications to the organic plots were according to the guidelines of the Organic Crop Improvement Association Inc. The compost was analyzed for total nitrogen and applied to provide 170 kg nitrogen per hectare for carrots and 300 kg nitrogen per hectare for cabbages, which assumed 50 percent availability of nitrogen. In addition to marketable yields, carrot leaves and roots and cabbage sections were digested and analyzed for 12 macro- and micronutrients. Vitamins C and E and α- and β-carotene of mature crops were determined. In 2 of the 3 years, vitamin C was also analyzed up to 24 weeks after harvest. Analysis of the 3 years of data showed that the yield and vitamin content of the carrots and cabbages were similar in organic and conventional crops.

The same authors (Warman and Havard, 1998) conducted a similar study on the yield, vitamin and mineral contents of organically and conventionally grown potatoes and sweet corn (maize) and obtained results very similar to those found with carrots and cabbages. Analysis of 3 years of data showed that the yield and vitamin C content of the potatoes were not affected by treatment. One possible reason for the finding that yield was similar on both treatments was that the amount of nitrogen applied to the soil was similar in both cases. However, the conventionally grown treatment out-produced the organically grown treatment for Pride and Joy (cv.) corn, but there was no difference between treatments in the yield of Sunnyvee (cv.) corn or in the vitamin C or E contents of the sweet corn kernels in any year.

Organoleptic quality

An important issue is whether consumers can detect any differences in the flavor or taste of organic produce. Anecdotally they can, but this is confounded with the fact that the organic produce is likely to be fresher. Scientific data on this issue are scarce.

Another complication in investigating a possible difference in flavor between organically and conventionally grown produce and fruit is that a comparison based on food items taken from supermarket shelves does not necessarily provide an accurate assessment of consumer preference. For instance, one type might have been stored longer than the other, affecting its freshness. Zhao *et al.* (2007) got around the problem by growing vegetable produce on replicated side-by-side plots to produce organic and conventional vegetables, then subjecting them to taste-testing at the same stage of freshness by a consumer panel of 100 members. The organically and conventionally grown products in each category were harvested at the same time and tested by the consumer panel the following day. The test involved red loose-leaf lettuce, spinach, arugula, mustard greens, tomatoes, cucumbers, and onions. The overall results showed no significant differences in consumer liking or consumer-perceived sensory quality between organically and conventionally grown vegetables (Table 3.11). The only exception was in tomatoes where the conventionally produced tomatoes were rated as having a significantly stronger flavor than the organically produced tomatoes. However, overall liking was the same for both organic and conventional samples. Since the conventionally grown tomatoes scored marginally higher in ripeness and a positive correlation was found between flavor intensity and ripeness, the flavor difference observed could not be simply ascribed to the contrasting growing conditions. This research adds weight to the argument that the superior flavor of organic produce reported by consumers is due to the food in question being fresher, especially when purchased at markets where the produce is sold quickly after harvesting. As Zhao *et al.* (2007) reported, when organically and conventionally produced vegetables were compared at the same stage of freshness, the taste was similar. The research was supported financially by the US Organic Farming Research Foundation.

A very useful feature of the test conducted by Zhao *et al.* (2007) is that consumers conducted the tests blind, i.e. with no information about the samples to be tasted. Labeling can affect the evaluation of organic and conventional products. Tests in which samples were labeled with the production method, or labeled with correct and false information, have shown that an organic label can lead to a significant increase in consumer rating (Schutz and Lorenz, 1976; von Alvensleben and Meier, 1990). Schutz and Lorenz (1976) found no significant difference in consumer acceptance

Table 3.11 Average taste panel scores for organically and conventionally produced vegetables

Vegetable	Production method	Overall liking[a]	Flavor intensity[b]
Lettuce	Organic	5.0	4.5
	Conventional	5.4	4.2
Spinach	Organic	6.3	3.8
	Conventional	6.3	3.7
Arugula	Organic	2.9	6.3
	Conventional	2.8	6.5
Mustard greens	Organic	2.7	6.6
	Conventional	3.0	6.5
Tomatoes	Organic	7.0	4.6
	Conventional	7.0	5.0
Cucumbers	Organic	7.1	4.1
	Conventional	7.0	4.0
Onions	Organic	6.7	5.7
	Conventional	6.7	5.9

No significant differences were found at $P < 0.05$ except for flavor intensity of tomatoes.
[a] On a 9-point hedonic scale, that is, 9 = like extremely, 1 = dislike extremely.
[b] On a 7-point scale, that is, 7 = extremely strong, bitter or sweet, 1 = barely any flavor or not bitter at all.
(*Source*: Zhao, X., Chambers, E. IV, Matta, Z., *et al.* (2007). Consumer sensory analysis of organically and conventionally grown vegetables. *Journal of Food Science*, **72**, S87–S91, Institute of Food Technologists, with permission of Wiley-Blackwell.)

of organically and conventionally grown vegetables, except when the samples were labeled and the ratings increased for the organically grown vegetables.

Gilsenan *et al.* (2010) conducted an experiment in Ireland to test whether there were any differences in the physicochemical and sensory properties of organic and conventional potatoes (cv. Orla). They found that conventional potatoes had a lower but significant dry-matter content and a slightly softer texture than the organic potatoes. However, a trained sensory panel did not find any significant differences between the conventional and organic potatoes (Table 3.12). The panel perceived the conventional baked potatoes to be slightly softer, less adhesive and wetter than the organic baked potatoes, but found no significant difference between the organic and conventional baked potato samples in terms of appearance, aroma, texture and taste acceptability. There was evidence of a slight preference for the conventional potatoes over the organic potatoes when baked. Based on their findings the researchers concluded that organic growing conditions affect the texture of raw and baked potatoes (cv. Orla), but do not appear to affect the appearance, taste or consumer acceptability of baked potatoes.

Table 3.12   Sensory evaluation scores of organic and conventional potatoes (cv. Orla) by a trained panel

Sensory parameters	Organic	Conventional	Significance of difference
Raw potatoes[1]			
External color	1.9[a]	2.0[a]	NS
Internal color	1.5[a]	1.6[a]	NS
Raw potato aroma	3.3[a]	2.9[a]	NS
Mustiness	2.7[a]	2.8[a]	NS
Earthiness	3.2[a]	3.6[a]	NS
Hardness	7.7[a]	7.8[a]	NS
Moistness	6.6[a]	6.4[a]	NS
Baked potatoes[1]			
Internal color	1.5[a]	1.4[a]	NS
Cooked potato aroma	6.4[a]	6.3[a]	NS
Mustiness	2.9[a]	2.9[a]	NS
Earthiness	3.9[a]	3.6[a]	NS
Hardness	3.5[a]	2.5[b]	$P < 0.05$
Adhesiveness	5.2[a]	4.3[b]	$P < 0.05$
Moistness	2.8[a]	3.9[b]	$P < 0.05$
Sweetness	3.4[a]	3.5[a]	NS
Aftertaste	3.2[a]	3.3[a]	NS
Appearance[2]	7.0[a]	7.1[a]	NS
Aroma[2]	7.2[a]	7.2[a]	NS
Texture[2]	6.8[a]	7.0[a]	NS
Taste[2]	7.0[a]	7.2[a]	NS

[1]Data are mean values of 30 organic and 30 conventional raw and cooked potatoes except where noted differently. Values bearing different superscripts are significantly different ($P < 0.05$). NS = not significant.
[2]Data are the mean values of 80 organic and 80 conventional baked potatoes. Values bearing different superscripts are significantly different ($P < 0.05$) for each attribute. NS = not significant.
(*Source*: Gilsenan, C., Burke, R.M. and Barry-Ryan, C. (2010). A study of the physicochemical and sensory properties of organic and conventional potatoes (*Solanum tuberosum*) before and after baking. *International Journal of Food Science and Technology*, **45**, 475–481, with permission of Wiley-Blackwell.)

Identification of organic produce

One topic of current interest to researchers is the use of chemical tests to try to identify a method of distinguishing between conventional and organic produce. One test that appears promising is the measurement of one of the stable isotopes of nitrogen, ^{15}N. As reported earlier in this chapter, one of the common effects of conventional production is an increase in the nitrogen (nitrate) content of produce. Camin *et al.* (2007) evaluated various markers for the traceability of potato tubers grown in an organic versus conventional regime. Identification of parameters separating organic and conventional produce should help prevent misconduct and could provide a firm basis for comparative assessment of the two types of produce. The researchers compared selected markers in organically and conventionally grown tubers in four separate field trials. Within each field trial, organic and conventional tubers were subjected to the same soil and climatic conditions since they were grown in adjacent plots. It was found that in all sites and in both cultivars tested, irrespective of environment, organic tubers showed a significant enrichment in ^{15}N content in comparison with tubers grown conventionally. Overall the average contents were 7.17 versus 3.36 percent, respectively. Other parameters tested, including ascorbic acid, protein content and dry matter, did not show the same consistent trends of variation throughout the four field trials. The researchers concluded that ^{15}N enrichment appears to be a promising discriminative marker for organic potatoes, possibly in combination with other markers.

Results obtained by Bagnaresi *et al.* (2007) confirmed that the most reliable indicator of organic production appears to be ^{15}N enrichment.

Kelly and Bateman (2010) reported additional findings on this topic, testing trace element and stable nitrogen isotope (^{15}N) concentrations in tomatoes and lettuce as possible parameters in distinguishing between conventional and organic cultivation. The results indicated that these data improved the correct classification of tomato samples but appeared to have had a limited effect on lettuce. The findings were also taken to support evidence suggesting systematic differences in the concentrations of certain elements such as manganese, calcium, copper, and zinc in crops grown organically and conventionally. The researchers postulated that the differences might be attributed to the presence of elevated levels of arbuscular mycorrhizal fungi (AMF) in soils cultivated organically.

Food from afar

One of the problems of living in northern countries is the difficulty in obtaining organic produce year-round. As a result, it is often necessary to buy foods that may have been shipped thousands of miles from countries with more favorable growing conditions. Since these foods are less fresh, they are likely to have less flavor and possibly a lower nutrient content than foods bought fresh from farmers' markets. Also, they may not have been grown under the organic conditions imagined by the consumer.

Another reason why some consumers are now refusing to buy organic foods that have been shipped from distant countries is the "carbon footprint" and the concept of sustainability. Organic farming seeks to promote the sustainability of the environment, with minimal influence on carbon emissions. As a result some consumers of organic foods now compromise and purchase organic produce when it can be obtained locally, and purchase regular produce when the organic supply is not available locally.

Finally: watch which salad veg you eat

It is now quite common to dish up exotic salads, sometimes using wild plants. These wild plants are natural after all, so why not?

Several media reports in the UK in 2008 highlighted the need for care in choosing what exotic greens to eat. The media, including the BBC (2008), recounted how UK celebrity chef Antony Worrall Thompson had recommended in a magazine that henbane would make a great addition to a summertime meal. Unfortunately, he had confused the plant with another of a similar name, fat hen. The problem with his recommendation is that henbane can be toxic and is linked historically to several famous poisonings, including the murder of the wife of Dr Crippen in 1910, and that of the King in Shakespeare's play *Hamlet*. The magazine and Mr Worrall Thompson subsequently issued an apology and a retraction of the recommendation. An ironical aspect of the incident is that the magazine in question is aimed at healthy living.

Conclusions

Levels of pesticides are likely to be lower in organic produce than in regular produce, but a reduced intake in our diet is probably not meaningful in terms of human health. Food-poisoning germs are a lot more important and can occur both in organic and regular foods.

Limited data, and these mainly on potatoes, suggest that there are few differences in nutritional content between organic and conventional produce and that the differences found are not of nutritional importance. Some studies have shown an increased content of vitamin C, β-carotene and phenolic compounds in organic produce, but a study with humans found that a high intake of these compounds from tomatoes did not lead to an increase in plasma concentration. A study with rats showed no difference in mineral retention when fed organic or conventional produce. Limited testing has been done to compare the taste and flavor of organic vs. conventional produce. Conventionally grown tomatoes were preferred to organic tomatoes but results suggest that, when vegetable produce is compared at the same stage of ripeness, no differences in taste or flavor are recorded.

References

Abreu, P., Relva, A., Matthew, S., *et al.* (2007). High-performance liquid chromatographic determination of glycoalkaloids in potatoes from conventional, integrated, and organic crop systems. *Food Control* **18**, 40–44.

Alvensleben, R. von and Meier, T. (1990). The influence of origin and variety on consumer perception. *Acta Horticulturae* **259**, 151–161.

Asami, D.K., Hong, Y.J., Barrett, D.M. and Mitchell, A.E. (2003). Comparison of the total phenolic and ascorbic acid content of freeze-dried and air-dried marionberry, strawberry, and corn grown using conventional, organic, and sustainable agricultural practices. *Journal of Agricultural and Food Chemistry* **51**, 1237–1241.

Bagnaresi, P., Camin, F., Moschella, A., *et al.* (2007). New analysis technique for identifying organic potatoes. *Informatore Agrario* **63**, 60–64.

Bailey, H., Zhao, P., Zhao, T. and Doyle, M.P. (1999). *Escherichia coli* and *Salmonella* on organic and conventional vegetables. Annual Report of Center for Food Safety and Quality Enhancement, University of Georgia 1999.

Baker, B.P., Benbrook, C.M., Groth, E. and Benbrook, K.L. (2002). Pesticide residues in conventional, integrated pest management (IPM)-grown and organic foods: insights from three US data sets. *Food Additives and Contaminants* **19**, 427–446.

BBC (2008). Chef sorry for poison plant error. British Broadcasting Corporation. http://news.bbc.co.uk/2/hi/uk_news/7540648.stm, accessed August 16, 2008.

Blaine, K. (2007). Food Safety and Organic Food. Food Safety Network, April 3, 2007. http://www.foodsafety.ksu.edu/en/article-details.php?a=2&c=6&sc=36&id=377, accessed April 10, 2007.

Bourn, D. and Prescott, J. (2002). A comparison of the nutritional value, sensory qualities, and food safety of organically and conventionally produced foods. *Critical Reviews in Food Science and Nutrition* **42**, 1–34.

Camin, F., Moschella, A., Miselli, F., *et al.* (2007). Evaluation of markers for the traceability of potato tubers grown in an organic versus conventional regime. *Journal of the Science of Food and Agriculture* **87**, 1330–1336.

Caris-Veyrat, C., Amiot, M.J., Tyssandier, V., *et al.* (2004). Influence of organic versus conventional agricultural practice on the antioxidant microconstituent content of tomatoes and derived purees; consequences on antioxidant plasma status in humans. *Journal of Agricultural and Food Chemistry* **52**, 6503–6509.

Chassy, A.W., Bui, L., Renaud, E.N.C., *et al.* (2006). Three-year comparison of the content of antioxidant microconstituents and several quality characteristics in organic and conventionally managed tomatoes and bell peppers. *Journal of Agricultural and Food Chemistry* **54**, 8244–8252.

Cieslak, P.R., Barrett, T.J., Griffin, P.M., *et al.* (1993). *Escherichia coli* 0157:H7 infection from a manured garden. *The Lancet* **342**, 367.

Claeys, W.L., Voghel, S. de, Schmit, J.F., *et al.* (2008). Exposure assessment of the Belgian population to pesticide residues through fruit and vegetable consumption. *Food Additives and Contaminants A* **25**, 851–863.

DiMatteo, K.T. (1997). Does organic gardening foster foodborne pathogens? Letter to the Editor. *Journal of the American Medical Association* **277**, 1679–1680.

Doyle, M.E. (2006). *Natural and Organic Foods: Safety Considerations. A Brief Review of the Literature*. FRI Briefings. Food Research Institute, University of Wisconsin–Madison, Madison, WI.

Forrer, G., Avery, A. and Carlisle, J. (2000). *Marketing and the Organic Food Industry*. National Center for Public Policy Research, Washington, DC.

Gambino, L., Graci, G., Gunnella, F., *et al.* (2007). Monitoring of agrochemical residues in horticultural produce: active substances in fruits and vegetables at the national average. *Informatore Agrario Supplemento* **63**, No. 39, Supplemento n. 1, 35–39.

Gilsenan, C., Burke, R.M. and Barry-Ryan, C. (2010). A study of the physico-chemical and sensory properties of organic and conventional potatoes (*Solanum tuberosum*) before and after baking. *International Journal of Food Science and Technology* **45**, 475–481.

Guadagnin, S.G., Rath, S. and Reyes, F.G.R. (2005). Evaluation of the nitrate content in leaf vegetables produced through different agricultural systems. *Food Additives and Contaminants* **22**, 1203–1208.

Haase, T., Schüler, C., Haase, N.U. and Hess, J. (2007). Suitability of organic potatoes for industrial processing: effect of agronomical measures on selected quality parameters at harvest and after storage. *Potato Research* **50**, 115–141.

Hajšlová, J., Schulzová, V., Slanina, P., *et al.* (2005). Quality of organically and conventionally grown potatoes: four-year study of micronutrients, metals, secondary metabolites, enzymic browning and organoleptic properties. *Food Additives and Contaminants* **22**, 514–534.

Hamilton-Miller, J.M. and Shah, S. (2001). Identity and antibiotic susceptibility of enterobacterial flora of salad vegetables. *International Journal of Antimicrobial Agents* **18**, 81–83.

Hartmann, S., Lacorn, M. and Steinhart, H. (1998). Natural occurrence of steroid hormones in food. *Food Chemistry* **62**, 7–20.

Hoefkens, C., Vandekinderen, I., Meulenaer, B. de, *et al.* (2009). A literature-based comparison of nutrient and contaminant contents between organic and conventional vegetables and potatoes. *British Food Journal* **111**, 1078–1097.

Hoogenboom, L.A.P., Bokhorst, J.G, Northolt, M.D., *et al.* (2008). Contaminants and microorganisms in Dutch organic food products: a comparison with conventional products. *Food Additives and Contaminants* **25**, 1197–1209.

Juday, D. (2000). Are organic foods really better for you? Natural grown killers in organic food make it no safer than produce grown in pesticides. *BridgeNews Service* (Knight Ridder) February, 14.

Karavoltsos, S., Sakellari, A., Dasenakis, M. and Scoullos, M. (2008). Cadmium and lead in organically produced foodstuffs from the Greek market. *Food Chemistry* **106**, 843–851.

Kelly, S.D. and Bateman, A.S. (2010). Comparison of mineral concentrations in commercially grown organic and conventional crops – tomatoes (*Lycopersicon esculentum*) and lettuces (*Lactuca sativa*). *Food Chemistry* **119**, 738–745.

Kristensen, M., Østergaard, L.F., Halekoh, U., *et al.* (2008). Effect of plant cultivation methods on content of major and trace elements in foodstuffs and retention in rats. *Journal of the Science of Food and Agriculture* **88**, 2161–2172.

Maggio, A., Carillo, P., Bulmetti, G.S., *et al.* (2008). Potato yield and metabolic profiling under conventional and organic farming. *European Journal of Agronomy* **28**, 343–350.

Magkos, F., Arvaniti, F. and Zampelas, A. (2006). Organic food: buying more safety or just peace of mind? A critical review of the literature. *Critical Reviews in Food Science and Nutrition* **46**, 23–56.

Marx, H., Gedek, B. and Kollarczik, B. (1994). Comparative studies of the bacterial and mycological status of ecologically and conventionally grown crops. *European Journal of Nutrition* **33**, 239–243.

McMahon, M.A.S. and Wilson, I.G. (2001). The occurrence of enteric pathogens and *Aeromonas* species in organic vegetables. *International Journal of Food Microbiology* **70**, 155–162.

Meerburg, B.G., Jacobs-Reitsma, W.F.J., Wagenaar, A. and Kijlstra, A. (2006). Presence of *Salmonella* and *Campylobacter* spp. in wild small mammals on organic farms. *Applied and Environmental Microbiology* **72**, 960–962.

Moreira, M. d.R., Roura, S.I. and Valle, C.E. del (2003). Quality of Swiss chard produced by conventional and organic methods. *LWT-Food Science and Technology* **36**, 135–141.

Mukherjee, A., Speh, D., Dyck, E. and Diez, G.F. (2004). Preharvest evaluation of coliforms, *Escherichia coli*, *Salmonella*, and *Escherichia coli* O157:H7 in organic and conventional produce grown by Minnesota farmers. *Journal of Food Protection* **67**, 894–900.

Mukherjee, A., Speh, D., Jones, A.T., *et al.* (2006). Longitudinal micro-biological survey of fresh produce grown by farmers in the upper Midwest. *Journal of Food Protection* **69**, 1928–1936.

NAS-NRC (1996). *Carcinogens and Anticarcinogens in the Human Diet: A Comparison of Naturally Occurring and Synthetic Substances*. National Research Council, National Academy Press, Washington, DC.

Nicholson, F.A., Hutchison, M.L., Smith, K.A., *et al.* (2000). *A Study on Farm Manure Applications to Agricultural Land and an Assessment of the Risks of Pathogen Transfer into the Food Chain*. Ministry of Agriculture, Fisheries and Food, London.

Phillips, C.A. and Harrison, M.A. (2005). Comparison of the microflora on organically and conventionally grown spring mix from a California processor. *Journal of Food Protection* **68**, 1143–1146.

Pest Management Regulatory Agency (PMRA) (2003). Pest Management Regulations. Health Canada Information Service, Ottawa, Ontario, Canada. www.hc-sc.gc.ca/pmra-arla/, accessed September 10, 2010.

Ponce, A.G., Roura, S.I., del Valle, C.E. and Fritz, R. (2002). Characterization of native microbial population of Swiss chard (*Beta vulgaris*, type cicla). *LWT-Food Science and Technology* **35**, 331–337.

Ponce, A.G., Roura, S.I., del Valle, C.E. and Fritz, R. (2003). Characterization of native microbial populations on Swiss chard (*Beta vulgaris*, type cicla) cultivated by organic methods. *LWT-Food Science and Technology* **36**, 183–188.

Raine, G. (2008). FDA to permit irradiation of spinach, lettuce. *San Francisco Chronicle*, August 22, 2008. http://articles.sfgate.com/2008-08-22/news/17123677_1_fda-irradiation-iceberg-lettuce, accessed March 23, 2009.

Rosenquist, H. and Hansen, A. (2000). The microbial stability of two bakery sourdoughs made from conventionally and organically grown rye. *Food Microbiology* **17**, 241–250.

Rossi, F., Godani, F., Bertuzzi, T., *et al.* (2008). Health-promoting substances and heavy metal content in tomatoes grown with different farming techniques. *European Journal of Nutrition* **47**, 266–272.

Russell, S. (2006). Spinach *E. coli* linked to cattle. Manure on pasture had same strain as bacteria in outbreak. *San Francisco Chronicle,* October 13, 2006. http://articles.sfgate.com/2006-10-13/news/17317271_1_dole-baby-spinach-spinach-field-spinach-outbreak, accessed March 23, 2009

Ruterberg, J. and Barringer, F. (2000). Apology highlights ABC reporter's contrarian image. *New York Times,* Aug. 14. p. C1. http://query.nytimes.com/gst/fullpage.html?, accessed September 5, 2007.

Schutz, H.G. and Lorenz, O.A. (1976). Consumer preferences for vegetables grown under "commercial" and "organic" conditions. *Journal of Food Science* **41**, 70–73.

Scott-Thomas, C. (2010). Black pepper producers reject salmonella allegations. *Food Navigator* February 11, 2010.

Szafirowska, A. and Elkner, K. (2008). Yielding and fruit quality of three sweet pepper cultivars from organic and conventional cultivation. *Vegetable Crops Research Bulletin* **69**, 135–143.

Szafirowska, A. and Elkner, K. (2009). The comparison of yielding and nutritive value of organic and conventional pepper fruits. *Vegetable Crops Research Bulletin* **71**, 111–121.

Tasiopoulou, S., Chiodini, A.M., Vellere, F. and Visentin, S. (2007). Results of the monitoring program of pesticide residues in organic food of plant origin in Lombardy (Italy). *Journal of Environmental Science and Health, B* **42**, 835–841.

Tauxe, R.V. (1997). Does organic gardening foster foodborne pathogens? In Reply. *Journal of the American Medical Association* **277**, 1679–1680.

Tschape, H., Prager, R. and Streckel, W. (1995). Verotoxinogenic *Citrobacter freundii* associated with severe gastroenteritis and cases of haemolytic uraemic syndrome in a nursery school: green butter as the infection source. *Epidemiology and Infection* **114**, 441–450.

UK Food Standards, Agency (2010). Organic Food. http://www.food.gov.uk/foodindustry/farmingfood/organicfood/, accessed September 2, 2010.

USDA (2007). US Department of Agriculture Pesticide Data Program, annual summary, calendar year 2006. USDA, Washington, DC.

USDA (2010). US Department of Agriculture Pesticide Data Program. http://www.ams.usda.gov/pdp, accessed April 26, 2010.

Warman, P.R. and Havard, K.A. (1997). Yield, vitamin and mineral contents of organically and conventionally grown carrots and cabbage. *Agriculture, Ecosystems and Environment* **61**, 155–162.

Warman, P.R. and Havard, K.A. (1998). Yield, vitamin and mineral contents of organically and conventionally grown potatoes and sweet corn. *Agriculture, Ecosystems and Environment* **68**, 207–216.

World Health Organization (2010). World Health Report – Promoting Fruit and Vegetable Consumption around the World. http://www.who.int/dietphysicalactivity/fruit/en/index2.html, accessed September 20, 2010.

Young, J.E., Zhao, X., Carey, E.E., *et al.* (2005). Phytochemical phenolics in organically grown vegetables. *Molecular Nutrition and Food Research* **49**, 1136–1142.

Yudelman, M., Ratta, A. and Nygaard, D. (1998). Pest management and food production: Looking to the future, 2020 Vision Brief 52 September, International Food Policy Research Institute. http://www.ifpri.org/2020/briefs/number52.htm, accessed August 27, 2009

Zhao, X., Chambers, E. IV, Matta, Z., *et al.* (2007). Consumer sensory analysis of organically and conventionally grown vegetables. *Journal of Food Science* **72**, S87–S91.

4 Fruit

Most consumers know that for optimal health they should include fruit in their diets. However, they have the same concerns about fruit as they have about vegetables: the possibility that chemical residues from sprays and other pesticides might endanger their health. As a result many consumers opt for organic fruit since it has been less subjected to spraying, and if they do buy conventionally grown fruit they wash it well or remove the skin before eating it.

Many consumers also purchase organic fruit with the expectation that it is more nutritious than conventional fruit.

Pesticide residues

Residues certainly can occur on fruit. For instance a US report found lower levels of pesticide residues in organic fruit than in regular fruit (Baker *et al.*, 2002). This was a collaborative study involving the Organic Materials Review Institute (Eugene, Oregon), Benbrook Consulting Services (Idaho) and the Consumers Union of the US (New York) and covered organically and conventionally grown fresh fruits and vegetables, as outlined in the previous chapter. The data analyzed were from three testing programs: the Pesticide Data Program of the US Department of Agriculture; the Marketplace Surveillance Program of the California Department of Pesticide Regulation; and private tests by the Consumers Union, an independent testing organization. The data sets were from 1989 to 1999.

The report found that organically grown foods consistently had about one-third as many residues as conventionally grown foods. In addition, conventionally grown samples were also more likely to contain multiple pesticide residues than were organically grown samples. Comparison of specific residues on specific crops found that residue concentrations in organic samples were consistently lower than in the conventional category, across all three data sets.

An example of the data presented in the report is shown in Table 4.1. These data indicated that during 1994–1999 organic fruit tested by the USDA was lower in pesticide residues than conventional fruit. As with

Organic Production and Food Quality, First Edition. Robert Blair.
© 2012 John Wiley & Sons, Ltd. Published 2012 by John Wiley & Sons, Ltd.

Table 4.1 Frequency of pesticide residues in fresh fruits by cultivation type, excluding the residues of banned organochlorines (USDA Pesticide Data Program Results, 1994–1999; from Baker *et al.*, 2002)

Fruit	Organic		Conventional	
	Number sampled	Percent positive	Number sampled	Percent positive
Apples	1	0	2294	94
Bananas	1	0	1134	58
Cantaloupe	3	33	1242	49
Grapes	4	25	1891	78
Oranges	7	14	1899	85
Peaches	2	50	1107	93
Pears	4	25	1777	95
Strawberries	8	25	1268	91
All fruit	30	23	12612	82

produce, this is a reassuring finding for the organic consumer. However, the amount of sampling in the organic fruit was much lower than in the conventional fruit, suggesting that more intensive sampling is needed and testing needs to be brought up to date.

Another issue from these findings is how the determined levels relate to the acceptable levels set out in the US Foods Regulations. In the largest commodity groups, fruits and vegetables, 44.2 percent and 73.8 percent of the samples, respectively, had no residues detected. Only 0.9 percent of the fruit samples and 2.4 percent of the vegetable samples contained violative residues.

The data collected during 2006 in the USDA Pesticide Monitoring Program show a more reassuring picture of the quality of conventional foods produced in the USA (Table 4.2). In that year 98.2 percent of all domestic foods analyzed were in compliance with established residue tolerances and formal action levels. Of the 1260 domestic samples tested, 68.8 percent had no detectable residues and 1.6 percent had violative residues. In the fruit group 55.8 percent of the samples had no detectable residues and 0.9 percent had violative residues.

These figures indicate that the monitoring and testing programs are having positive effects for the consumer. However, they indicate a rate of testing that many consumers would like to see expanded. Also, no distinction is drawn between conventional and organic foods.

The corresponding results for 2007 (USDA, 2008) indicate a continued reduction in residue violations in fruit. Overall, the level of residues detected (the number of residues detected divided by the total number of analyses performed for each commodity) was 1.9 percent. Over 99 percent of the samples analyzed did not contain residues above the

Table 4.2 Frequency of pesticide residues in US domestic fruit, excluding the residues of banned organochlorines (USDA, 2007: http://www.ams.usda.gov/pdp)

Product	Total samples analyzed	Percent positive	Samples in violation	Violations over tolerance	Violations no tolerance
Blackberries	2	100	0	0	0
Blueberries	16	81.2	0	0	0
Boysenberries	0	0	0	0	0
Cranberries	16	62.5	0	0	0
Grapes, raisins	1	100	0	0	0
Raspberries	4	50.0	0	0	0
Strawberries	38	44.7	0	0	0
Grapefruit	1	0.0	0	0	0
Lemons	2	100	0	0	0
Oranges	21	42.9	0	0	0
Other citrus fruit	4	25.0	0	0	0
Apples	123	34.1	0.8	0	1
Pears	15	46.7	0	0	0
Other core fruit	0	0	0	0	0
Apricots	1	0.0	100	0	1
Avocadoes	2	100	0	0	0
Cherries	6	33.3	0	0	0
Nectarines	6	16.7	16.7	0	1
Peaches	48	20.8	0	0	0
Plums	1	100	0	0	0
Olives	1	100	0	0	0
Bananas, plantains	5	100	0	0	0
Cantaloupe	4	50	0	0	0
Watermelon	3	100	0	0	0
Other melons	3	33.3	0	0	0
Other fruits	7	57.1	0	0	0
Apple juice	10	100	0	0	0
Orange juice	1	100	0	0	0
Other fruit juices	3	100	0	0	0
Fruit jams/jellies/ pastes/toppings	0	0	0	0	0
Total	344	44.2	0.9	0	3

safety limits (tolerances) established by the US Environmental Protection Agency (EPA), and 96.7 percent of the samples analyzed did not contain residues for pesticides that had no tolerance established. Of the 12 689 total samples collected and analyzed, 9734 were fruit and vegetable commodities. The percentage of total residue detections for fresh fruit and vegetables ranged from 0.8 to 3.8 percent, with a mean of 2.2 percent. The percentage of total residue detections for processed fruit and vegetables ranged from 0.6 to 2.2 percent, with a mean of 1.3 percent.

A total of 49 fruit and vegetable samples was found to have residues at levels exceeding the established tolerance, including 6 blueberry samples, 5 nectarine samples and 3 peach samples. Of those 49 samples, 8 were reported as imported produce. Blueberries were found to have 1/504 violations to a fungicide (sample imported from Chile), 3/504 violations to an insecticide and 2/711 to a herbicide. Nectarines were found to have 1/563 violations to an insecticide (sample imported from Chile) and 4/563 to another insecticide. Peaches showed 1/555 violations to a fungicide and 2/555 to an insecticide (samples imported from Chile). As in previous years no comparable data on organic fruit were presented.

As stated in Chapter 3, the USDA tested 13 381 samples in 2008 (USDA, 2010). The total included 10 382 samples of fruit and vegetables. The tests showed that, for fresh and processed fruit and vegetables, almonds, honey, catfish, and rice, approximately 76.4 percent of all samples tested were from US sources, 19.8 percent were imports, 2.7 percent were of mixed national origin, and 1.1 percent were of unknown origin. Of the 11 960 samples analyzed, the overall percentage of total residue detections was 1.6 percent. The percentage of total residue detections for fresh fruit and vegetables ranged from 0 to 3.3 percent, with an average of 1.9 percent. The percentage of total residue detections for processed fruit and vegetables ranged from 0 to 2.2 percent, with an average of 0.6 percent. Residues exceeding the tolerance were detected in 0.5 percent of the 11 960 samples tested in 2008: 58 samples with one residue exceeding and two samples with two residues exceeding the established tolerance. Residues with no established tolerance were found in 3.7 percent of the samples. In most cases, these residues were detected at very low levels and some residues may have resulted from spray drift or crop rotations.

An example of the change in residue levels in fruit can be seen with peaches (Table 4.3). Pesticides detected in peaches in 2001–2002 are different from those detected in 2008–2009, due to EPA's ongoing review of data to verify safety of pesticides used in food production to ensure protection of human health and the environment. The pesticide monitoring program re-tests high-consumption foods every 5 years or sooner so that the EPA has pertinent and current knowledge of residues in foods. When comparing pesticide residue data on peaches over the time period 2002 to 2008, significant changes in crop protection practices are evident. Residues of some of the organophosphates, such as azinphos-methyl, decreased from 46.5 percent to 25.6 percent of the samples tested, chlorpyrifos residues decreased from 35.5 to 17.2 percent, and phosmet residues decreased from 64.8 to 30.7 percent. For other older pesticides such as carbaryl, the change in residue level was from 32.3 to 9.7 percent and for iprodione from 54.7 to

Table 4.3 Comparison of residues of pesticides in peaches during 2002 and 2008 (USDA Pesticide Data Program Results, 2010. http://www.ams.usda.gov/pdp)

Pesticide	Peaches 2002				Peaches 2008			
	Minimum concentration detected (ppm)	Maximum concentration detected (ppm)	EPA tolerance (ppm)	Samples with detections (%)*	Minimum concentration detected (ppm)	Maximum concentration detected (ppm)	EPA tolerance (ppm)	Samples with detections (%)*
Azinphos methyl	0.005	0.52	2.0	46.5	0.005	0.29	2.0	25.6
Boscalid		NA			0.002	0.48	1.7	12.5
Carbaryl	0.002	3.2	10	32.3	0.002	1.9	10	9.7
Chlorpyrifos	0.002	0.079	0.05	34.5	0.003	0.11	0.1	17.2
Cyhalothrin, lambda		NA			0.005	0.032	0.5	22.8
Esfenvalerate	0.006	0.033	10.0	12.7	0.025	0.07	10.0	29.8
Esfenvalerate + Fenvalerate total	0.006	0.35	10.0	11.6	0.0023	0.084	10.0	37.4
Fenvalerate	0.009	0.044	10.0	13.4	ND	ND	10.00	0
Fludioxonil	0.020	1.8	5.0	30.6	0.003	4.7	5	47.6
Formetanate HCl		NA			0.0002	1.1	5.0	19.3
Iprodione	0.014	33	20	54.8	0.025	12	20	31.2
Methamidophos	0.001	0.49	0.02	12.6	0.002	0.023	0.02	1.5
Methoxyfenozide		NA			0.002	0.18	3.0	14.9
o-Phenylphenol	0.005	0.046	20	7.3	0.017	0.038	20	16.3
Phosmet	0.002	1.4	10	64.8	0.005	0.82	10	30.7
Propiconazole	0.024	0.085	1.0	6.2	0.060	0.51	2.0	11.5
Spinosad A		NA			0.003	0.026	0.2	19.4

* = Only pesticides with residue detections in at least 10 percent of samples for either year are shown. NA = Not analyzed. No samples were analyzed for that pesticide/commodity pair. ND = Non-detects. Samples were analyzed for that pesticide/commodity pair, but none had a residue.

31.2 percent. These pesticides have been replaced by safer pesticides such as boscalid, methoxyfenozide, and spinosad A.

Fresh fruits and vegetables are often eaten in a fresh, raw state. Health experts and the US Food and Drug Administration advise that washing fresh fruit and vegetables before eating is a healthful habit. Consumers can reduce and often eliminate pesticide residues present by washing fruits and vegetables with cool or lukewarm tap water.

Although it seems logical that lower levels of pesticide residues should be found in organic foods, a study in New Zealand found very little evidence that organic foods do contain lower levels. This was after a review of 209 relevant reports in the international scientific literature (Bourn and Prescott, 2002). Consequently, the New Zealand researchers recommended that further analytical studies needed to be conducted to clarify the issue.

A 2005 joint Israeli–Belgian review of the scientific evidence entitled "Need for research to support consumer confidence in the growing organic food market" (Siderer *et al.*, 2005) came to a similar conclusion. After reviewing the relevant scientific reports on a comprehensive range of foods (cereals and cereal products, potatoes, vegetables and vegetable products, fruit and fruit products, wine, beer, bread, milk and dairy products, meat and egg products, eggs and honey), the authors concluded that there has been very little published information on residue levels, insufficient to allow clear conclusions to be drawn. In the case of conventionally cultivated produce they pointed out that the European Commission had published the findings of the 2001 pesticide residue monitoring program (European Commission, 2003). The report compiled EU-wide analyses of pesticide residues in 46 000 samples of fruits, vegetables and cereals. The residue levels were found to fall within a range that would not cause harm if ingested: Fifty-nine percent of the samples contained no detectable residues while 37 percent of the samples contained detectable residues at or below the maximum residue level (MRL). The MRL in the EU pesticide regulations is the maximum amount of pesticide expected in a foodstuff that has been produced according to good agricultural practice. It is not a safety limit. On average, 3.9 percent of samples exceeded the MRL, ranging from 1.3 to 9.1 percent in different member states.

The UK Pesticide Residues Committee in 2010 reported results on the sampling of fruit purchased at the retail level in 2009. This organization tested 1848 samples of fruit and vegetables for up to 278 pesticides, a total of over 427 000 food and pesticide combinations. Residues were found in 1105 samples (59.8 percent), and 40 samples (2.2 percent) contained residues above the MRL (Table 4.4). In the fruits category, residues above the MRL were found in grapes, melons and pears. There was evidence that the incidence was higher in products imported into the UK, making follow-up more difficult to ensure compliance with approved production practices at the farm level. The report did not specify how residues above the MRL related to the acceptable upper level of safety for the pesticides in

Table 4.4   Results of pesticide sampling of fruits in the UK

Fruit	Number of samples	Number of samples with residues at or below the MRL*	Number of samples with residues above the MRL*	Number of samples with more than one residue
Apples	144	131	0	112
Bananas	96	84	0	56
Grapes	152	134	3	111
Grapefruit	54	54	0	54
Lemons	90	80	0	72
Melons	96	63	1	37
Limes	36	34	0	16
Pears	148	137	4	118

*MRL = Maximum residue level permitted in the regulations.
(*Source*: UK Pesticide Residues Committee (2010). Report for 2009. © Crown Copyright material reproduced with the permission of the Controller, HMSO.)

question. Also, the report did not provide information on pesticide levels in organic fruit.

Between 2004 and 2007, 40 to 49 samples of fruit, vegetables and orange juice of stated organic origin were analyzed for pesticide residues in Ireland (Irish National Food Residue Database, 2009). Between three (2005) and eight (2006) samples contained measurable pesticide residues and, in one case, the level determined exceeded the appropriate MRL. While the levels of pesticide residues determined were relatively low and did not pose a concern for the consumer, their occurrence in products labeled as organic suggests that the production systems in place did not conform with organic production requirements.

In spite of the deficiency in the amount of recorded data, a research study conducted in Greece showed that pesticide residues in olive oil could be reduced by organic cultivation (Tsatsakis *et al.*, 2003). The olive oil samples were collected from Crete during 1997–1999 and tested for two pesticides, fenthion and dimethoate. It was found that the average concentrations of fenthion in conventional olive oils were 0.1222, 0.145 and 0.1702 mg/kg, and for dimethoate were 0.0226, 0.0264 and 0.0271 mg/kg for 1997, 1998 and 1999, respectively. The comparable figures for fenthion in organic olive oils were 0.0215, 0.0099 and 0.0035 mg/kg, and for dimethoate were 0.0098, 0.0038 and 0.0010 mg/kg, respectively.

This research showed that the organic olive oils contained significantly lower concentrations of the two pesticides. The researchers pointed out, however, that all of the samples contained residue levels lower than the maximum residue levels (MRLs) adopted by the Food and Agriculture Organization of the United Nations, the World Health Organization and the Codex Alimentarius.

Therefore, while it is apparent that there needs to be better documentation of pesticide levels in conventional foods before it can be stated conclusively that organic foods are better in this respect, the work on olive oil indicates that residues in at least some types of fruit can be reduced by organic cultivation methods.

Other risks with fruit

Chemical residues

Chemical contaminants resulting from pollution of the environment can occasionally be found in both organic and conventional food products (Magkos *et al.*, 2003). Some of these contaminants may pose health risks because of their persistence in the soil, particularly in countries such as The Netherlands which have had a long history of intensive agriculture. Among these contaminants are chlorinated hydrocarbons, polychlorinated biphenyls and some heavy metals. Areas of high contamination may occur due to adjacent or previous industrial activity. Therefore the presence or absence and the concentration of these toxic agents in food, both organic and conventional, depend mainly on farm location. Similar levels of these pollutants have been reported in organic and conventional foods (Lecerf, 1995; Woese *et al.*, 1997; Kumpulainen, 2001; Worthington, 2001).

A current concern is the possibility of food contamination with cadmium. This element is regarded as an environmental contaminant and classified as a human carcinogen. It is commonly regarded as entering the human food chain through food, water, sewage sludge, environmental contamination and cigarette smoking. There is a possibility that the use of fertilizer in growing food crops may increase the human intake of this element. Use of fertilizer is banned or restricted in organic farming, although organic farmers may still be permitted to use crude phosphate rock containing variable amounts of cadmium. Regular phosphate fertilizer is likely to have been cleaned of most cadmium before application to conventional crops (Kirchmann and Thorvaldsson, 2000).

Trace element accumulation, particularly of cadmium, has been reported to be higher in the soil of organic farmland (Moolenaar, 1999). Increased concentrations of metals in soils, however, do not necessarily result in increased metal contents in the plants grown in these soils or in the produce (Moolenaar and Lexmond, 1999). Although comparative analyses of organic and conventional foods are limited, the few studies published to date show no consistent difference in cadmium levels (Lecerf, 1995; Woese *et al.*, 1997; Jorhem and Slanina, 2000; Worthington, 2001; Malmauret *et al.*, 2002). As shown in Chapter 6 on meats, the concentration of cadmium was found to be higher in organic pork than in conventional pork in one study (Lindén *et al.*, 2001).

Other data indicate a higher concentration of cadmium in conventional foods than in organic foods (Karavoltsos *et al.*, 2002), attributed to the presence of cadmium in the fertilizer used. In the fruits category, peaches recorded the lowest content (0.1–0.3 ng/g fresh-weight basis) and melons the highest at 5.6–5.9 ng/g. In a further study the same research group (Karavoltsos *et al.*, 2008) reported values less than 1.0 ng/g fresh-weight basis in organic apples, kiwi, oranges, mandarins, apricots and peaches. This report did not contain comparable values for conventional fruit; however, it did contain data on analyzed lead content, since this is another toxic element. Lead content (ng/g fresh-weight basis) ranged from less than 8.0 in peaches to 27.2 in oranges. No comparable values for lead in conventional fruit were presented.

Copper is another element that has come under scrutiny because of the use of Bordeaux mixture (an aqueous solution of copper sulfate) and other copper salts as fungicides in organic agriculture. It has been suggested that organic growers apply more frequent treatments with copper-based fungicides on their crops than do conventional farmers (Trewavas, 2001). Despite the fact that their use was banned after 2002, copper salts remain persistent in soil and thus raise concern over food contamination with copper. A review of several studies indicated that organic fruits, vegetables and grains contain on average approximately 10 percent more copper than conventional produce (Worthington, 2001). However, only the percentage difference was considered and no actual contents were presented. Therefore, whether this finding implies a beneficial or hazardous effect on human health remains at present unclear, especially since dietary requirements and upper tolerable levels of copper intake are still subject to conjecture (Buttriss and Hughes, 2002).

Residues of heavy metals can occur in fruit juice, as was evidenced by the recall of pear juice in March 2008 in Canada. The problem was arsenic contamination, affecting two brands of juice distributed by Loblaws Inc. One of the two products involved was organic (President's Choice Organics Pear Juice from Concentrate for Toddlers) and was, unfortunately, intended for children.

The recall by the Canadian Food Inspection Agency and Loblaws Inc. was conducted as a precaution to prevent long-term exposure to arsenic. According to Health Canada the levels found, while higher than would normally be found in this type of juice, were not high enough to represent a risk to children or adults from short-term exposure.

"Arsenic is a toxic heavy metal that may be carcinogenic and may pose developmental risks to children," the Canadian Food Inspection Agency stated in its news release.

One welcome aspect of the recall is that it occurred as a result of vigilance on the part of the official food surveillance activities of government agencies. The exact source of the contamination was not reported. No illnesses were recorded as a result of the contamination of the products recalled.

This incident demonstrated how the change in the organic industry to adjust to increasing consumer demand affected the recall. The food distributor had to conduct a nationwide product recall throughout Canada, and in order to determine the source of the problem had to contact suppliers and producers in Turkey, New Jersey and North Carolina.

There do not appear to be other documented cases of chemical contamination of organic fruit, or of comparable studies involving organic and conventional fruit.

Recent information on the contamination of fruit with chemicals and pesticides suggests that the concerns held by the average consumer may be somewhat exaggerated, at least in some of the developed countries. For instance, tests on locally produced food showed that the average New Zealand diet presents no chemical-residue food safety concerns (New Zealand Food Safety Authority, 2009). This program tests more than 120 commonly eaten foods to estimate the average dietary exposure to chemical residues, contaminants and selected nutrients. Foods tested for the current study were split into two groups: those from the regions and those available nationally. In the first quarter, 61 regional foods were tested from supermarkets and shops in Auckland, Napier, Christchurch and Dunedin in January and February 2009. They were prepared ready for consumption, then sent to a laboratory in Hamilton for analysis (on an as-received moisture basis). While the survey was not intended to be a compliance survey, any issues of non-compliance with allowable limits for residues or contaminants are acted upon. The results demonstrated the high quality of food in New Zealand, with no problems from residues. From 60 000 analyses, there were just two findings that required follow-up. Neither involved fruit. One was a non-compliance issue in tomatoes from Napier, where the pesticide azaconazole was found at slightly above the maximum residue limit of 0.05 mg/kg. The New Zealand maximum residue limits are not safety limits, which are in most cases several times higher, but rather an indicator of good agricultural practice (i.e. similar to the European MRLs). The concentration found in the tomatoes did not pose a food safety or health concern, but it did require that the producer follow good agricultural practice. The other was higher than expected levels of lead in breads from Napier. Again, the levels did not pose a health concern.

Microbial problems

One study reported similar concentrations of coliform bacteria on conventional and organic fruits and vegetables (Doyle, 2006). No other documented reports appear to be available.

Mycotoxins

There are recorded instances of risks from mycotoxins in fruit juices unless precautionary steps are taken. One of these mycotoxins is patulin. More frequent and higher concentrations of patulin have been reported in organic fruit juices and purées than in conventional products (Piemontese *et al.*, 2005).

A group at the Department of Food Safety and Food Quality in Belgium studied this mycotoxin, which can be produced by fungi in spoiled apples. Levels of patulin were measured in organic, conventional and homemade apple juice marketed in Belgium (Baert *et al.*, 2006). In all, 177 apple juice samples were analyzed: 65 organic, 90 conventional, and 22 homemade. Patulin was detected in 22 samples (12 percent), and the amount could be measured in 10 (6 percent). The patulin content was higher than the European legal limit of 50 µg/liter in two samples of organic apple juice. Overall, the incidence of patulin in organic (12 percent), conventional (13 percent), and homemade (10 percent) apple juices was not significantly different. However, the average concentration of patulin in contaminated samples was significantly higher in organic (43.1 µg/liter) than in conventional (10.2 µg/liter) or homemade (10.5 µg/liter) juice. The highest patulin concentrations were found in the most expensive apple juices. This relationship was not observed with the conventional apple juices.

These findings indicate that consumers purchasing apples should avoid any spoiled fruit, also that organic producers need to observe strict quality control of their produce. Windfall apples are best avoided altogether by producers and consumers.

In addition, pasteurized apple juice is recommended over raw juice. In 1996 the FDA announced that Odwalla was recalling all of the company's juice products that contained unpasteurized apple juice (Food & Drug Administration, 1996). The recall followed 13 reported cases of *E. coli* O157: H7 illness that had been linked to the company's unpasteurized apple juice by the Seattle-King (Washington) County Department of Public Health. When the outbreak was over, one child was dead from complications arising from *E. coli* O157:H7 infection, and more than 65 individuals were confirmed infected with the bacterium in the western United States and British Columbia (Canada). Of these reported cases, more than a dozen developed hemolytic uremic syndrome (HUS), a life-threatening condition that causes the body's major organs, particularly the kidneys, to fail. As a direct result of the outbreak, Odwalla began pasteurizing its juices. The outbreak also resulted in a federal government requirement requiring warning labels to be placed on all unpasteurized fruit and vegetable juice containers.

Cloning and gene-modified fruit

The public appears to have no concerns about the cloning of food-producing plants, either because they approve of the technology or are unaware of its use. Taking a cutting from a plant to multiply it rather than growing it from seed is cloning, although that may not be apparent to the grower or the consumer. A good example of the widespread use of cloning is apple production. Most or all of the apples in stores or grown in our gardens are from cloned trees. This is because it is known that apple trees grown from seed do not breed true. So, for decades, growers have attached a shoot of the desired fruit tree to the rootstock of another tree in a process called grafting. Once the graft "takes", the resulting tree is an exact genetic copy (a clone) of the desired fruit tree. Making a genetic copy of the preferred fruit tree in this way is the only method of ensuring reliable and consistent apple quality. The process is repeated every time more fruit trees are required.

Genetic modification (GM) does not appear to be used in the breeding of fruit trees, or at least not to the extent that it is used in vegetable production (e.g. corn). Therefore it is not a current concern of the consumer.

Nutrient concentrations

A strong perception held by many consumers is that organic production results in fruit of higher nutritional value. For instance, Kihlberg and Risvik (2007) found that the majority of organic consumers think that organic food tastes better than conventional food. However, data on this issue are very limited (Doyle, 2006). Also it is not clear from some reports whether the fruit composition was compared on a similar moisture basis. As with vegetable produce, there is evidence that some organic fruit is drier than conventionally grown fruit. Unless this factor is taken into account, a higher content of a nutrient might be explained by a higher dry-matter content (lower moisture). A slightly drier fruit may also have a more intense flavor due to the higher concentration of nutrients, and as a result may be preferred by the consumer. As with vegetable produce, other factors that need to be comparable in an assessment (and explanation) of differences in nutritional concentration between organic and conventional fruit include total yield, stage of maturity at harvest, cultivar, strain and variety.

The effect of differing dry-matter content on nutrient level can be seen in the results of Cayuela et al. (1997). Strawberries grown organically had higher dry-matter, sugar and vitamin C contents than a similar variety of strawberries grown conventionally, when the results were expressed on a fresh-weight basis. However, when these data are converted to a dry-weight basis the nutrient contents are similar or slightly higher in the conventional fruit (Table 4.5).

The higher dry-matter content of the organic strawberries was associated with a higher organoleptic quality (detailed in a subsequent section of this chapter).

An experiment conducted in Jordan under plastic-house conditions yielded similar findings. Strawberries (cv. Camaroza) produced conventionally were larger and had a higher moisture content than organically produced fruit

Table 4.5 Effect of conventional and organic production on the nutrient content of strawberries (from Cayuela *et al.*, 1997; © American Chemical Society)

Determined content (mean of 11 harvests)	Conventional fresh-wt basis	Conventional dry-wt basis	Organic fresh-wt basis	Organic dry-wt basis
Dry matter %	8.1	–	8.9	–
Sugars °Brix	7.5	9.3	8.4	9.4
Vitamin C mg/liter	700	864	720	809

(Abu-Zahra and Tahboub, 2009). On the other hand, fruit produced organically had a more intense color and higher dry-matter, total phenol, crude fiber, carotene and vitamin C contents than fruit produced conventionally. Conventionally produced fruit tended to have a higher total titratable acidity content.

The effects of organic and conventional nutrient amendments on the yield and quality of strawberries was reported by Hargreaves *et al.* (2008). In this study treatment did not affect sugar content or total antioxidant capacity. Conventional fertilizer treatment resulted in an increased sulfur and manganese content of berries compared with the organic treatment (dry-matter basis). The potassium and phosphorus contents of berries differed between years.

Other available data on the effect of cultivation system on the vitamin C content of fruits have yielded controversial results (Woese *et al.*, 1997). The results are difficult to interpret because the growing conditions varied widely and different methods of sampling and analysis were used. There are also species differences. Vitamin C content was increased in organic as compared with conventional peaches, but not in pears (Bergamo and Cappelloni, 2002).

Andrews *et al.* (2001) investigated the nutrient content of apples (Golden Delicious) grown under organic, conventional, and integrated production systems in Washington State, USA. They found that after 5 years a lower available soil nitrogen in the organic system was associated with significantly lower fruit tissue nitrogen. Fruit calcium contents rose consistently in all three production systems over 4 cropping years. In addition, there were significant differences in fruit nutrient ratios among the three systems. Fruit cortical nitrogen concentration was lowest in the organic production system in both 1998 and 1999 (Table 4.6). There were no differences in fruit phosphorus among systems in 1998, and in fruit potassium, magnesium, and calcium in either year, although fruit calcium had doubled since 1995. Phosphorus concentration was lowest in the organic fruit in 1999. Fruit boron concentration was lower in the organic system than in the conventional system in 1998, and in the organic and integrated systems in 1999. Ratios of fruit cortical nutrient contents were calculated. In 1998, the N:Ca ratio was higher in the integrated production system (9.8) than in either the

Table 4.6 Concentrations of minerals in Golden Delicious apples grown under organic, conventional and integrated systems (Andrews et al., 2001)

Mineral (%) dry-wt basis	Organic	Conventional	Integrated
1998			
N	0.40b	0.46ab	0.54a
Ca	0.06a	0.07a	0.06a
P	0.09a	0.10a	0.10a
K	0.82a	0.90a	0.85a
Mg	0.04a	0.05a	0.04a
B	5.0b	7.0ab	7.2a
1999			
N	0.30b	0.40a	0.41a
Ca	0.10a	0.10a	0.09a
P	0.072b	0.080a	0.075ab
K	0.80a	0.88a	0.88a
Mg	0.04a	0.05a	0.05a
B	7.8b	11.5a	7.5b

*a b c. Means followed by different letters in a row are significantly different at $P < 0.05$.

organic (6.5) or conventional (6.5) systems. In 1999, the N:Ca ratios among the production systems were not different; however, the N:P ratio was higher in the conventional (5.1) and integrated (5.4) systems than in the organic (4.2) system. Other fruit cortical nutrient ratios, such as, Mg:Ca, K:Ca, and Mg + K:Ca, were not different among the systems in either year.

Organic fruit was firmer at harvest and immediately after removal from either regular (RA) or controlled (CA) atmosphere storage than fruit grown conventionally (Table 4.7). There were no differences in fruit firmness among the three systems after 7 days in air at 20°C following removal from either storage regime. Similar differences in fruit firmness between the systems and storage regimes were found in 1998. In 1999, organic fruit was significantly firmer than conventional fruit after 6 months' CA storage, either immediately after removal from storage or after 7 days in air at 20°C. Fruit from the organic system had higher soluble solids than the conventional fruit either immediately after removal from 3 months of CA storage or after 7 days in air at 20°C (Table 4.7). Organic fruit also had lower acidity than the conventional fruit, except immediately after removal from CA storage. The higher soluble solids and lower acidity in organic fruit resulted in higher soluble solids:acidity ratios than either conventional or integrated fruit. A sensory panel confirmed the differences in sweetness versus tartness of organic and conventional fruit but did not find any firmness differences.

Table 4.7 Fruit firmness, soluble solids content, and titratable acidity of organic, conventional, and integrated apples at harvest and after storage at 1°C in regular or controlled atmosphere for 3 months (Andrews *et al.*, 2001)

Parameter (units)	Organic	Conventional	Integrated
Harvest			
Firmness (n)	70.6a	68.5b	65.4c
Soluble solids (%)	13.3a	13.2a	13.4a
Acidity (% malic acid equivalents)	0.76b	0.81a	0.74b
Regular atmosphere			
Firmness (n)	50.2a	48.0b	49.0ab
Soluble solids (%)	14.6a	14.6a	14.3a
Acidity (% malic acid equivalents)	0.52b	0.58a	0.57a
Controlled atmosphere			
Firmness (n)	60.7a	59.0b	57.0c
Soluble solids (%)	15.3a	14.7b	14.7b
Acidity (% malic acid equivalents)	0.70a	0.71a	0.70a
Harvest + 7 days			
Firmness (n)	66.9a	65.5ab	64.8b
Soluble solids (%)	14.6a	14.0b	14.6a
Acidity (% malic acid equivalents)	0.71b	0.74a	0.75a
Regular atmosphere + 7 days			
Firmness (n)	48.1a	48.3a	47.8a
Soluble solids (%)	14.7a	15.0a	14.7a
Acidity (% malic acid equivalents)	0.47b	0.50a	0.51a
Controlled atmosphere + 7 days			
Firmness (n)	57.0a	57.0a	57.3a
Soluble solids (%)	15.6a	15.0b	15.2ab
Acidity (% malic acid equivalents)	0.59b	0.62a	0.61a

*a, b, c. Means followed by different letters in a row are significantly different at $P < 0.05$.

Lombardi-Boccia *et al.* (2004) investigated the concentration of antioxidant vitamins (vitamin C, vitamin E, β-carotene) and phenolics (total polyphenols, phenolic acids, flavonols) in yellow plums (*Prunus domestica* L) grown conventionally and organically on the same farm. Conventional plums were grown on tilled soil. Three organic cultivations were performed: tilled soil, soil covered with *Trifolium*, and soil covered with natural meadow. Differences in macronutrients were marginal, whereas the concentration of antioxidant vitamins and phenolic compounds differed markedly among cultivations (Table 4.8 and Table 4.9). Concentrations of vitamin C, α- and γ-tocopherols, and β-carotene were higher in organic plums grown on soil covered with natural meadow. The highest content of phenolic acids was detected in plums grown on soil covered with

Table 4.8 Nutrient composition of conventionally and organically grown plums (from Lombardi-Boccia *et al.*, 2004, © American Chemical Society; mean values fresh-weight basis)

Composition	Conventional Tilled soil	Organic		
		Tilled soil	*Trifolium*	Meadow
Edible portion (%)	95.5	96.8	96.6	96.9
Water (g/100 g)	88.7	88.7	88.1	88.1
Ash (g/100 g)	0.33	0.37	0.37	0.37
Crude protein (g/100 g)	0.51	0.51	0.53	0.51
Lipid (g/100 g)	0.13	0.13	0.13	0.13
Minerals				
P (mg/100 g)	11.9	14.6	15.6	15.1
Na (mg/100 g)	1.46	1.0	1.2	1.3
K (mg/100 g)	174	201	218	207
Mg (mg/100 g)	4.90	5.80	5.70	5.40
Ca (mg/100 g)	4.16	4.85	4.49	4.26
Fe (mg/100 g)	0.29	0.27	0.26	0.27
Zn (μg/100 g)	60	70	40	90
Cu (μg/100 g)	60	50	60	60
Mn (mg/100 g)	50	50	50	40
Dietary fiber				
Total fiber (g/100 g)	1.38	1.36	1.50	1.41
Soluble fraction (g/100 g)	0.56	0.55	0.61	0.56
Insoluble fraction (g/100 g)	0.82	0.82	0.89	0.80
Carbohydrates				
Fructose (g/100 g)	1.08	1.38	1.21	1.35
Glucose (g/100 g)	1.97	2.11	2.00	2.00
Sucrose (g/100 g)	2.23	1.82	2.75	2.50
Sorbitol (g/100 g)	1.16	0.96	1.18	1.20
Total sugar (g/100 g)	6.44	6.27	7.14	7.05
Organic acids				
Citric acid (mg/100 g)	27.0	25.7	25.8	25.2
Malic acid (g/100 g)	1.98	1.98	2.02	2.02
Vitamins				
Vitamin C (mg/100 g)	2.00	1.60	2.10	2.20
α-Tocopherol (μg/100 g)	446	415	411	585
γ-Tocopherol (μg/100 g)	7.2	9.7	9.7	11.0
β-Carotene (μg/100 g)	107	68	77	117
Vitamin K1 (mg/100 g)	12.4	10.9	12.6	9.7

Table 4.9 Content of total polyphenols, phenolic acids, and flavonols in conventionally and organically grown plums (from Lombardi-Boccia et al., 2004, © American Chemical Society; mean values, fresh-weight basis)

Composition	Conventional Tilled soil	Organic Tilled soil
Total polyphenols (mg tannic acid/100 g)	121	88
Phenolic acids (mg/kg)		
Protocatechuic	0.8	0.6
Caffeic	20.6	22.6
trans-p-Coumaric	8.5	8.9
Ferulic	8.0	9.3
Chlorogenic	25.2	37.5
Neo-chlorogenic	52.0	46.0
Flavonols (mg/kg)		
Myricetin	0.9	1.1
Quercetin	30.2	19.6
Kaempferol	0.6	1.7

Trifolium. The content of total polyphenols was higher in conventional plums. Quercetin was higher in conventional plums, but myricetin and kaempferol were higher in organic plums. With the same cultivar and climatic conditions, the type of soil management was found to be of primary importance in influencing the concentration of these compounds.

Organic and conventional Hayward kiwifruits grown on the same farm and harvested at the same maturity stage were compared in a California study (Amodio et al., 2007). Quality parameters included morphological (shape index) and physical (peel characteristics) attributes. Maturity indices (CO_2 and C_2H_4 production, firmness, color, soluble-solids content and acidity) and content of compounds associated with flavor and nutritional quality (minerals, sugars and organic acids, vitamin C, total phenolics, and antioxidant activity) were determined at 0, 35, 72, 90 and 120 days of storage at $0°C$. They were also determined after 1 week of shelf-life simulation at $20°C$, after each storage duration.

Organically and conventionally grown kiwifruits had similar soluble solids content at harvest, but conventional kiwifruits had a higher firmness and L^* value, and a lower hue angle and chromaticity, resulting in a lighter green color when compared with the organic kiwifruits. These differences were maintained for all the storage periods, with the soluble-solids content increasing more in conventionally grown kiwifruits. The two production systems resulted in different morphological attributes since organic kiwifruits exhibited a larger total and columella area, smaller flesh area, more spherical shape, and thicker skin compared with conventional kiwifruits. All the main mineral constituents were more concentrated in organic kiwifruits, which also had higher levels of vitamin C and total phenol content, resulting in a higher antioxidant activity. Composition of sugars and organic acids was not affected by production system.

Researchers at the USDA Genetic Improvement of Fruit and Vegetable Laboratory and Rutgers University tested blueberries grown on five organic and conventional farms in New Jersey that had the same soil, weather and harvesting conditions (Wang *et al.*, 2008). The effect of cultivation practices on fruit quality and antioxidant capacity in a Bluecrop variety of highbush blueberries (*Vaccinium corymbosum* L.) was evaluated. Results from this study showed that blueberries grown organically had a significantly higher content of sugar (fructose and glucose), malic acid, total phenolics and total anthocyanins, and a higher antioxidant activity (oxidant radical absorbance capacity, ORAC) than fruit grown conventionally. The organic blueberries contained about 50 percent higher levels of total anthocyanins, the natural plant phytochemicals that give blueberries their dark color. They also had 67 percent more total phenolics. In addition the organic fruit had higher contents of myricetin 3-arabinoside, quercetin 3-glucoside, delphinidin 3-galactoside, delphinidin 3-glucoside, delphinidin 3-arabinoside, petunidin 3-galactoside, petunidin 3-glucoside, and malvidin 3-arabinoside than conventional fruit. The results were expressed on a fresh-weight basis and these researchers did not report whether the moisture content varied in the berries grown organically and conventionally.

Tarozzi *et al.* (2006) studied the content of phytochemicals, total antioxidant activity and *in vitro* bioactivity in red oranges grown conventionally or organically. Red oranges ('Tarocco' cultivar) harvested from certified producers operating in eastern Sicily and using either conventional (integrated) or organic production methods were purchased from a single retail outlet (all the integrated/organic fruits were individually labeled as products produced under controlled cultivation conditions in line with the provisions of statutory European Community regulations regarding integrated and organic farming). Four different lots of six fruits at comparable periods from harvest time were analyzed. All fruits were selected to be similar in size and weight. No physical defects or signs of pathogen contamination were apparent. All fruits were processed within 48–72 hours of purchase. Results showed that organic red oranges had a higher phytochemical content (i.e. phenolics, anthocyanins and vitamin C), total antioxidant activity and bioactivity than conventional red oranges (Table 4.10). In addition, the results indicated that red oranges had a strong capacity to inhibit the production of conjugated diene-containing lipids and free radicals in rat cardiomyocytes and differentiated Caco-2 cells. The authors did not state whether the results were expressed on a similar dry-matter basis.

This finding is consistent with changes in the antioxidant system of plants grown organically. Phenolic compounds are thought to be involved in the

Table 4.10   Contents of phenolics, vitamin C and anthocyanins in the edible tissue of fresh red oranges

Production type	Phenolics (mg gallic acid equivalents/100 g)	Vitamin C (mg/100 g)	Anthocyanins (mg/100 g)
Conventional	74.0	45.0	5.56
Organic	79.0	53.0	6.65

(*Source*: Tarozzi, A., Morroni, F., Cantelli-Forti, G., *et al.* (2006). Antioxidant effectiveness of organically and non-organically grown red oranges in cell culture systems. *European Journal of Nutrition* **45**, 152–158. Reprinted with kind permission of Springer Science & Business Media.)

defense mechanism of plants by acting as a chemical barrier to invading phytopathogens, and modulation of the levels of these compounds by pesticides (herbicides, insecticides and fungicides) has been observed in some plants (Daniel *et al.*, 1999; Brandt and Mølgaard, 2001).

Several investigations have involved a study of the sugar content of fruit, because of its importance to diabetics, e.g. Hecke *et al.* (2006). Statistical analyses of two groups of apples grown organically or by conventional (integrated) cultivation methods showed no significant differences in total sugar content, but a clear tendency to higher phenol and malic acid contents in organically grown cultivars. This study found considerable cultivar differences and it is unfortunate that the same cultivars were not used at the conventional and organic sites. As a result it is not possible to draw a clear distinction between the effects of conventional and organic production. The study does, however, provide useful data on cultivar differences.

Total sugar content of the cultivars Gala, Elstar, Idared, Golden Delicious, Braeburn and Fuji ranged between 115 and 150 g/kg fresh weight, but the cultivar Jonagold had a higher sugar concentration (183 g/kg fresh weight). Fructose and sucrose were the major sugar components. Jonagold and Fuji had higher fructose concentrations than Braeburn and Golden Delicious, followed by Idared, Elstar and Gala. The sucrose concentration of Jonagold was about 85 g/kg higher than in the other cultivars. Glucose concentration was between 5 and 20 g/kg, ranging in order from Braeburn, Fuji, Idared, Gala, Jonagold and Elstar to Golden Delicious. The highest sorbitol concentration was measured at about 6 g/kg in Fuji, Jonagold, Braeburn and Elstar.

Cultivars from organically grown sites were found to have a broader range of total sugar concentration. Three varieties (Steirische Schafnase, Steirischer Maschanzker and Lavantaler Bananenapfel) showed a sugar content of 200 g/kg fresh weight. The lowest sugar content was found in Baumanns Rentte and Gravensteiner. The content of sugar in the other varieties ranged between 115 and 160 g/kg fresh weight, similar to the values in cultivars from integrated production. As with the conventional fruit, sucrose and fructose were the main sugar components in the organically grown cultivars. A high glucose content was found in Roter Boskoop. Cox Orange and Danziger Kantapfel showed a particularly high sorbitol content.

Most of the organically grown cultivars contained total acid concentrations similar to those of the ones grown under integrated cultivation, with malic acid being the main acid component. The total acid concentration of all cultivars ranged between 6 g/kg fresh weight (Golden Delicious) and 14 g/kg fresh weight (Elstar), with malic acid being the main acid found. Citric acid concentrations were found in Fuji, Jonagold, Idared and Golden Delicious, but could not be detected in Braeburn, Elstar or Gala owing to technical difficulty. Shikimic acid concentrations were highest in Gala, but did not reach a value above 0.05 g/kg in any cultivar. Very low concentrations of fumaric acid were found (0.5–1.9 mg/kg) in all cultivars except Elstar, which analyzed at up to 10 times the concentration found in the other cultivars. Noticeably higher acid concentrations were found in the cultivars Steirische Schafnase, Roter Boskoop and Steirischer Maschanzker. Citric acid was found in Baumanns Renette, Cox Orange and Gravensteiner.

The phenolic concentration of the cultivars grown under integrated conditions was lowest in Golden Delicious (1334 µg/ml). Braeburn, Gala, Fuji, Jonagold and Idared showed similar concentrations of phenolic compounds, which ranged between 2142 and 2642 µg/ml. Some organically grown cultivars showed much higher concentrations of phenols. Steirischer Maschanzker had values higher than 3000 µg/ml; Lavanttaler Bananenapfel, Roter Boskoop, Weiszliger Klarapfel and Steirische Schafnase showed concentrations even higher than 6000 µg/ml. Lower levels of phenols were found in London Pepping, Cox Orange and Danziger Kantapfel.

The results of this investigation agreed with general assumptions that 100 g fruit contains 12 g carbohydrates. Most of the cultivars from conventional production had this concentration of carbohydrate. However, Jonagold had higher values. Also, many of the regional, organically grown apple varieties did not fit the general assumption, some having higher sugar values (Steirische Schafnase, Lavanttaler Bananenapfel, Steirischer Maschanzker) while others had lower sugar values (Baumanns Renette, Danziger Kantapfel and Gravensteiner).

Similar findings on apple quality were reported by Roussos and Gasparatos (2009). In this study two apple (Bork) orchards, a conventional and an organic one, were compared in terms of plant growth, marketable fruit quality attributes (fruit weight, shape, color, phenolic compound concentration, nutrients) and yield. The two orchards were located nearby, in order to exclude possible pedoclimatic influences on the measured variables. The two management systems resulted in similar fruit quality attributes in terms of total soluble solids, juice pH, titratable acidity and color indexes. The conventionally grown trees produced almost twice the yield of the organically managed ones. The flesh plus peel portion of the conventionally produced fruits exhibited higher total flavonoid and *o*-diphenols concentration, while the flesh portion presented higher flavonoid concentration. Nitrogen concentration was higher in all portions of conventionally grown fruits, whereas potassium, calcium, sodium and manganese concentrations were higher in the flesh portion of organically produced fruits.

The phenolic content of apples has been investigated by other researchers. Eleven organically grown apple cultivars and 11 apple cultivars from conventional (integrated) production in Austria and Slovenia were analyzed for their content of phenolic compounds in peel and pulp (Veberic *et al.*, 2005). Chlorogenic acid, *p*-coumaric acid, procyanidin B3, protocatechuic acid, (−)-epicatechin, phloridzin, rutin and quercetin-3-rhamnoside were identified in apple peel. In apple pulp, (+)-catechin was also identified in all cultivars. Some other phenols (procyanidin B3, rutin and quercetin-3-rhamnoside) could not be identified or were not properly separated. No differences were found in the phenolic content of apple peel between organic and conventional cultivars. Organic apples, however, had a higher content of phenolic substances in the pulp than conventional cultivars. The researchers hypothesized that the difference might be due to the source of genotype or to the production method, and also that higher concentrations of phenolic compounds in organically grown cultivars could be a result of plant response to stress. The apple peel was found to contain higher concentrations of identified phenols than the pulp. However, since the apple peel represents up to 10 percent of the whole fruit, it was suggested that the phenolic compounds in the pulp are of greater importance to the consumer than those in the peel.

Grapes and wine contain large amounts of phenolic compounds, and moderate wine consumption is believed to reduce the risk of coronary heart disease. The mechanism is considered to be that phenolic compounds present in red wines cause an increase in serum total antioxidant capacity, and thereby inhibit low-density lipoprotein (LDL) oxidation. Among fruit grapes, must and pomace are a valuable source of phenolic antioxidants. A study by Yildirim *et al.* (2005) measured the content of phenols and the antioxidant potential in organically grown grapes, pomace (the solid remains after grapes are pressed for juice), must (fresh-pressed grape juice), and wines. The findings indicated that the total phenolic content and its capacity to increase antioxidant activity (AOA) and inhibit LDL were highest in pomace, grape and must, respectively. For wines, similar results were found in red wines made from Cabernet Sauvignon, Merlot and Zinfandel, respectively. Based on grape color and the total phenolic content, it was calculated that the antioxidant potential was 46 percent or more for red wines and only 3 to 6 percent for white wines. This study also reported a positive correlation between total phenolic content and antioxidant activity.

Another study examined the levels of anthocyanin in the skin of Syrah grapes (Vian *et al.*, 2006). Anthocyanins are responsible for the red, purple and blue colors of many fruits, vegetables, grains and flowers, and are also of interest for possible health benefits as antioxidants. The grapes were harvested at different stages of ripening, from either organic or conventional systems. Samples of grapes were collected from the veraison stage (the time they began to change color) to full maturity. Using high-performance liquid chromatography, the researchers measured the contents of nine different anthocyanins. The total content of anthocyanins in conventionally grown grapes was significantly higher than in grapes grown organically. The accumulation of anthocyanins reached a maximum 28 days after veraison (also the period of highest temperature), and then decreased until harvest. In all samples, grapes from conventional plots had higher proportions of delphinidin, petunidin, malvidin and acylated malvidin glucosides compared with grapes from organic plots. For each grape variety considered, climatic conditions and agronomic practices were important factors. The researchers speculated that the higher levels of stress during the dry and hot summer of 2002 probably resulted in the higher levels of anthocyanin in the conventionally grown grapes as a result of a higher stress from application of synthetic chemicals in those plots.

It is clear from the available data that organic and conventional fruits do not differ significantly in their content of major nutrients. On the other hand, higher contents of phenolic compounds and antioxidants have been reported in some organic fruit. Where a higher content of nutrients has been reported in organic fruit, it is clear that the difference may be explained by a higher dry-matter content. Researchers working in this area need to take this factor into account in reporting their findings. Another factor that needs to be reported is the total yield of fruit, since the organic plants may produce the same amount of nutrients but in a reduced harvest. In addition, the stage of maturity at harvest, and cultivar, strain or variety investigated, need to be reported. All of this information would help to provide a scientific explanation for any differences found in nutrient content.

Some of the studies discussed above illustrate another factor that needs to be taken into account in comparisons of nutrient content. Differences in dry-matter content of major and trace elements in plant foodstuffs between harvest years can exceed the differences between production methods. Therefore several harvests have to be taken before any overall conclusion can be reached.

For instance, the influence of harvest year was investigated in a study involving the quality of extra-virgin olive oils (EVOO) from organic and conventional farming (Ninfali *et al.*, 2008). The oils were extracted from Leccino and Frantoio olive (*Olea europaea*) cultivars, grown in the same geographical area of Italy under either organic or conventional methods. Extra-virgin olive oils were produced with the same technology, and samples were analyzed for nutritional and quality parameters. Sensory evaluation was also conducted by a trained panel. Significant differences were found in these parameters between organic and conventional oils in some years, but no consistent trends across the three years were found.

Sensory analysis showed only slight differences. The results showed that organic versus conventional cultivation did not affect the quality of EVOO consistently, at least in the parameters measured. Genotype and year-to-year changes in climate, instead, had more marked effects.

Plants grown organically are known to produce higher levels of secondary metabolites such as polyphenolic compounds, which act as natural pesticides. Whether an increase in the content of these compounds in organic food is beneficial to the human consumer is not yet clear.

Appearance and organoleptic qualities

Another important aspect of organic fruit for the consumer is whether it looks different and whether it has a better flavor than conventional fruit. For the grower an important aspect is the keeping quality, apart from the appeal to the consumer.

Strawberries grown organically were found to have higher dry-matter, sugar and vitamin C contents than a similar variety of strawberries grown conventionally, when the results were expressed on a fresh-weight basis (Table 4.5, Cayuela et al., 1997). The higher dry-matter content of the organic strawberries was associated with a higher organoleptic quality (Table 4.11). Organically grown fruits had a higher resistance to deterioration during

Table 4.11   Effect of conventional and organic production on the eating and retail qualities of strawberries (Cayuela et al., 1997; © American Chemical Society)

Determined content (mean of 11 harvests)	Conventional fresh-wt basis	Organic fresh-wt basis	Significance of difference*
Color (scale 0–5)	3.2	3.6	**
Brightness (scale 0–5)	3.0	3.3	**
Appearance (scale 0–5)	3.2	3.5	**
Odor (scale 0–5)	1.8	2.5	**
Hardness (scale 0–5)	2.6	2.5	ns
Juiciness (scale 0–5)	2.9	3.3	***
Acidity (scale 0–5)	2.8	2.3	***
Sweetness (scale 0–5)	1.8	2.6	***
Taste (scale 0–5)	2.1	2.9	***
Overall grade (scale 1–9)	5.3	6.5	***
Commercial losses (%: fruit with 33% or more with surface damage or rot)			
Day 2	0	0	ns
Day 5	4.8	1.4	*
Day 11	65.7	35.0	***
Day 13	89.6	22.6	***

*ns = Nonsignificant. *, ** and *** significant at the 0.05, 0.01 and 0.001 levels respectively.

simulated marketing conditions, and thus better keeping quality. These benefits were associated with a smaller crop, yield of the conventionally grown strawberries being 76 percent greater.

These findings suggest that organic production may have a greater beneficial effect on the consumer and retail quality of soft fruit than is the case with tree fruits.

Lester *et al.* (2007) investigated the quality of organic and conventionally grown Rio Red grapefruit fruit and juice. These researchers pointed out that most claims for organic produce being better tasting and more nutritious than conventional produce are largely unsubstantiated. They concluded that this is due mainly to a lack of rigor in research studies to ensure similar production variables of both production systems, such as microclimate, soil type, fertilizer elemental concentration, previous crop, irrigation source and application, plant age, and cultivar. The research in question was designed to address these issues with Texas commercially grown conventional and certified organic Rio Red (red-fruited) grapefruit. Whole grapefruits from each production system were harvested between 08.00 and 10.00 at commercial early (November), mid (January), and late (March) season harvest periods for three consecutive years. Within each harvest season, conventional and organic whole fruits were compared for market qualities (fruit weight, specific gravity, peel thickness, and peel color), and juices were compared for market qualities (specific gravity, percentage juice, and color), human health-bioactive compounds (minerals, vitamin C, lycopene, sugars, pectin, phenols, and nitrates), and consumer taste intensity and overall acceptance. Conventional fruit was better colored and higher in lycopene, and the juice was less tart, lower in the bitter principle naringin, and better accepted by the consumer panel than the organic fruit. Organic fruit had a commercially preferred thinner peel, and the juice was higher in vitamin C and sugars and lower in nitrate and the drug interactive furanocoumarins.

Reganold *et al.* (2001) assessed the production of Golden Delicious apples grown under conventional, integrated and organic systems by measuring fruit yields, size and grade; tree growth; leaf and fruit mineral contents; fruit maturity; and consumer taste preference. Cumulative yields were similar but organic fruit was smaller. Fruit tissue nutrient analyses indicated some inconsistent differences. Mechanical analysis of fruit firmness at harvest and after storage in 1998 and 1999 showed that organic fruit was firmer than or as firm as conventional and integrated fruit. Ratios of soluble solids to acidity were most often highest in organic fruit (the report does not specify whether these analyses were conducted on a similar dry-matter basis). These data were confirmed in taste tests by untrained sensory panels which found the organic apples to be sweeter after 6 months of storage than conventional apples and less tart at harvest and after 6 months of storage than conventional and integrated apples. The same taste tests, however, could not discern any difference in firmness among apples in the three systems at harvest or after storage. Taste tests also indicated that the integrated apples had a better flavor after 6 months' storage but found no differences among organic, conventional and integrated apples in texture or overall acceptance. In spite of the reported benefits it was calculated that, without price premiums for organic fruit, the conventional system resulted in the highest net return to the grower, followed by the integrated system, and then the organic system.

The quality and flavor of organically and conventionally grown apples, peaches, bananas, strawberries, apricots and melons were compared in an Italian study (Capitini *et al.*, 2002). All of the fruit was bought in the same supermarket, and the fruit was evaluated by a taste panel for appearance, color, firmness and flavor. The findings were not clear cut. Conventionally grown bananas were given higher scores for firmness and appearance than organic bananas, but organic bananas and apricots received higher scores for consistency than those grown conventionally. Organic peaches received a significantly higher score for brightness than conventionally grown peaches, and the color of organic melons and apricots was deemed significantly superior to that of conventionally grown fruit. The color of conventionally grown apples was preferred to that of organic apples, and organic strawberries were found superior to conventionally grown strawberries with regard to firmness and color. Organic bananas, apricots and strawberries were given significantly higher scores for flavor than those grown conventionally, but there were no significant differences between the flavor scores of organic and conventionally grown peaches, apples and melons.

Róth *et al.* (2007) observed no significant difference in terms of aroma volatiles and other quality attributes of apples coming from different regions or from conventional and organic production systems either at harvest or after storage. The apples (cv. Jonagold) were from three different regions of Belgium. In each region one organic and one integrated orchard with identical climatic and soil characteristics was sampled. The fruit was stored in air and under controlled atmospheric conditions (CA: 1% O_2, 2.5% CO_2) at $1°C$ for 6 months. The acoustic stiffness, firmness, soluble solids content, acid and sugar contents, and the aroma profile were measured. Quality parameters were analyzed right after harvest and storage. At both times an additional shelf-life experiment was carried out to simulate the conditions in the commercial chain. There was a considerable softening during storage in air and in shelf-life, but not under CA conditions. Immediately after harvest, high malic acid, quinic acid and sucrose contents were observed, while glucose and citric acid contents were higher after storage. The aroma profile changed in shelf-life, except for apples stored in air, which even immediately after storage already had an aroma profile comparable to that after shelf-life. The volatile responsible for the typical apple aroma (2-methylbutyl acetate) had the highest relative abundance after CA storage and subsequent shelf-life, followed by apples immediately after CA storage. The researchers concluded that storage conditions generally had a greater influence on quality of the apples than production system.

Reig *et al.* (2007) conducted a study in Spain to compare the quality of organically and conventionally grown apples (*Malus domestica* cv. Fuji and Golden Delicious). In spite of a slight increase in vitamin C in organic Fuji apples, no differences in total antioxidant activity were found between organically and conventionally grown apples at harvest. In contrast, organic Fuji apples showed higher values of firmness, acidity, soluble solids content, and L* and a* values but lower weight values. Similarly, the organically grown Golden Delicious apples exhibited the same increases but only when the fruits were picked later. At both harvest dates, organic Golden Delicious apples were significantly less mature (lower starch index), but not the Fuji apples. These results showed that organic production may delay tree-fruit ripening and also improve the fruit eating quality.

A similar investigation was conducted by Amarante *et al.* (2008). The objective of the study was to assess the yield and quality of apples produced under conventional and organic production systems in Southern Brazil. The orchards consisted of alternate rows of 10- to 12-year-old Royal Gala and Fuji apple trees on M.7 rootstocks. Eighteen apple trees of each cultivar and management system were randomly selected for study during two growing seasons (2002/2003 and 2003/2004). The organic management system resulted in lower concentrations of potassium, magnesium and nitrogen in leaves and fruits, smaller fruits for both cultivars, and lower fruit yield for Fuji than from the conventional production system (Table 4.12). For both cultivars, fruits from the organic orchard harvested at commercial maturity had a more yellowish skin background color, higher percentage of blush in the fruit skin, higher soluble solids content, higher density, higher flesh firmness, and higher severity of russet than fruits from the conventional orchard. Fruit from the organic orchard had lower titratable acidity in Royal Gala, and a higher incidence of moldy core and lower incidence of water-core in Fuji than fruit from the conventional orchard. In spite of these differences a sensory panel found no significant differences in taste, flavor or texture between fruit from the production systems for either cultivar. These results are consistent with those of DeEll and Prange (1992), showing that trained sensory panelists were unable to detect any difference between organic and conventional apples.

Preserves

Scientists at the Institute of Food Research in the UK (Gunning *et al.*, 2009) have reported findings indicating another aspect of the importance of fruit in our diets. The findings relate to pectin, a fiber component found in fruits and vegetables. The laboratory study found that a fragment released from pectin binds to galectin 3 (Gal3), a protein involved in cancer. This gives

Table 4.12   Mean fruit quality of Royal Gala and Fuji apples under a conventional and an organic production system at commercial harvest maturity. Values represent the average of two growing seasons, 2002/2003 and 2003/2004

Fruit quality attribute	Conventional orchard	Organic orchard	Significance of difference
Royal Gala			
Fruit weight (g)	118.3	83.3	***
Flesh firmness (N)	95.3	100.6	**
Soluble solids content (°Brix)	11.3	12.0	**
Starch index (1–5)	2.6	2.5	ns
Titratable acidity (%)	0.5	0.4	*
Skin background color (1–8)	4.7	5.5	**
Blush in the fruit skin (%)	76.0	82.7	**
Fruit density (g/cm^3)	0.867	0.886	***
Russet severity (cm^2/fruit)	2.3	4.8	***
Taste (1–5)	3.2	3.1	ns
Flavor (1–5)	2.8	3.1	ns
Texture (1–5)	3.7	3.8	ns
Fuji			
Fruit weight (g)	135.4	91.5	***
Flesh firmness (N)	81.2	88.0	***
Soluble solids content (°Brix)	12.8	13.4	**
Starch index (1–5)	3.7	3.8	ns
Titratable acidity (%)	0.4	0.4	ns
Skin background color (1–8)	4.1	5.1	***
Blush in the fruit skin (%)	82.2	91.1	***
Fruit density (g/cm^3)	0.873	0.889	***
Russet severity (cm^2/fruit)	5.2	6.9	***
Watercore (%)	50.6	40.1	**
Moldy core (%)	5.4	21.9	***
Taste (1–5)	3.5	3.3	ns
Flavor (1–5)	3.2	2.9	ns
Texture (1–5)	3.8	3.7	ns

*ns = Nonsignificant. *, ** and ***significant at the 0.05, 0.01 and 0.001 levels respectively. (*Source*: Amarante, C.V.T., André Steffens, C.T., Mafra, A.L. and Albuquerque, J.A. (2008). Yield and fruit quality of apple from conventional and organic production systems. *Revista Pesquisa Agropecuária Brasileira* **43**, 333–340. Reprinted with permission of Embrapa Informacao Tecnologica.)

the pectin fragment its anticancer effect. These findings correlate with population studies carried out by groups such as the European Prospective Investigation of Cancer (EPIC), which found a strong link between dietary fiber and a reduced risk of cancers of the gastrointestinal tract. However, the exact mechanism for the protective effect of fiber could not be identified in the European studies. The new research supports the earlier findings of scientists at the University of Georgia, who found that 40 percent of prostate cancer cells died after exposure to pectin powder or heat-treated citrus pectin. Jackson *et al.* (2007) found in a laboratory study that a commercially available pectin powder induced apoptosis (death) in human androgen-independent prostate cancer cells. Other related studies on rats and cell cultures suggest that pectin also fights lung and colon cancers. Applications of these findings could include the development of pectin-based pharmaceuticals, nutraceuticals, and also recommended diet changes aimed at combating prostate cancer occurrence and progression.

Fruits rich in pectin include cooking apples, damsons, gooseberries, bitter oranges, lemons, blackcurrants and redcurrants. Blackberries, plums, apricots and raspberries are also good sources. Pectin is also found in jams and fruit preserves.

No detailed information appears to be available on the levels of pectin in organic vs. conventional fruit. Lester *et al.* (2007) reported no significant difference in the pectin content of juice from conventional and organic grapefruit.

Conclusions

A consumer concern is that fruit might contain chemical residues from sprays and other pesticides that could endanger their health and that organic fruit would be more healthful. Although it seems logical that lower levels of chemical pesticide residues should be found in organic fruits, there is little documented evidence in the scientific literature or in the results of governmental food monitoring programs to prove this point convincingly. Also, the available findings indicate that the low levels of residues reported in conventional foods rarely exceed acceptable limits.

The findings indicate that consumers purchasing apples should avoid any fruit containing rot, also that organic producers need to observe strict quality control of their produce. Windfall apples are best avoided altogether by producer and consumer, to avoid documented problems related to mold contamination.

It is clear from the available data that organic and conventional fruits do not differ significantly in their content of major nutrients. On the other hand, higher contents of phenolic compounds and antioxidants have been reported in some organic fruit, which might be beneficial in terms of human health. Where a higher content of nutrients has been reported in organic fruit, it is apparent that the difference may be explained by a higher dry matter content.

Another important consumer aspect of organic fruit is whether it looks different and whether it has a better flavor than conventional fruit.

The available evidence indicates that no clear conclusion can be drawn due to differing results being reported for several fruits and an inconsistency in the findings over several harvests. Some organic fruit may have a more intense flavor because of a higher dry-matter content. The higher dry-matter content can be beneficial in soft fruit such as strawberries because it improves the keeping quality.

References

Abu-Zahra, T.R. and Tahboub, A.A. (2009). Strawberry (*Fragaria* × *Ananassa* Duch) fruit quality grown under different organic matter sources in a plastic house at Humrat Al-Sahen. *Acta Horticulturae* **807**, 353–358.

Amarante, C.V.T., André Steffens, C.T., Mafra, A.L. and Albuquerque, J.A. (2008). Yield and fruit quality of apple from conventional and organic production systems. *Revista Pesquisa Agropecuária Brasileira* **43**, 333–340.

Amodio, M.L., Colelli, G., Hasey, J.K. and Kader, A.A. (2007). A comparative study of composition and postharvest performance of organically and conventionally grown kiwifruits. *Journal of the Science of Food and Agriculture* **87**, 1228–1236.

Andrews, P.K., Fellman, J.K., Glover, J.D. and Reganold, J.P. (2001). Soil and plant mineral nutrition and fruit quality under organic, conventional, and integrated apple production systems in Washington State, USA. *Acta Horticulturae* **564**, 291–298.

Baert, K., De Meulenaer, B., Kamala, A., *et al.* (2006). Occurrence of patulin in organic, conventional, and handicraft apple juices marketed in Belgium. *Journal of Food Protection* **69**, 1371–1378.

Baker, B.P., Benbrook, C.M., Groth, E. and Benbrook, K.L. (2002). Pesticide residues in conventional, integrated pest management (IPM)-grown and organic foods: insights from three US data sets. *Food Additives and Contaminants* **19**, 427–446.

Bergamo, P. and Cappelloni, M. (2002). Modulation of antioxidant compounds in organic vs. conventional fruit (peach, *Prunus persica* L, and pear, *Pyrus communis* L.). *Journal of Agricultural and Food Chemistry* **50**, 5458–5462.

Bourn, D. and Prescott, J. (2002). A comparison of the nutritional value, sensory qualities, and food safety of organically and conventionally produced foods. *Critical Reviews in Food Science and Nutrition* **42**, 1–34.

Brandt, K. and Mølgaard, J. P. (2001). Organic agriculture: does it enhance or reduce the nutritional value of plant foods? *Journal of the Science of Food and Agriculture* **81**, 924–931.

Buttriss, J. and Hughes, J. (2002). A review of the MAFF Optimal Nutrition Status research programme: folate, iron and copper. *Public Health Nutrition* **5**, 595–612.

Capitini, R., Massantini, R. and Mencarelli, F. (2002). A comparison of the quality and organoleptic traits in fruit produced by organic and conventional methods. [Confronto qualitativo ed organolettico di frutti da agricoltura biologica e convenzionale.] *Industrie Alimentari* **41**, 789–794.

Cayuela, J.A., Vidueira, M., Albi, M.A. and Gutiérrez, F. (1997). Influence of the ecological cultivation of strawberries (*Fragaria × Ananassa* cv. Chandler) on the quality of the fruit and on their capacity for conservation. *Journal of Agricultural and Food Chemistry* **45**, 1736–1740.

Daniel, O., Meier, M.S., Schlatter, J. and Frischknecht, P. (1999). Selected phenolic compounds in cultivated plants: ecologic functions, health implications, and modulation by pesticides. *Environmental Health Perspectives* **107**, 109–115.

DeEll, J.R. and Prange, R.K. (1992). Postharvest quality and sensory attributes of organically and conventionally grown apples. *HortScience* **27**, 1096–1099.

Doyle, M.E. (2006). *Natural and Organic Foods: Safety Considerations. A Brief Review of the Literature.* FRI Briefings. Food Research Institute, University of Wisconsin–Madison, Madison, WI.

European Commission (2003). Monitoring of Pesticide Residues in Products of Plant Origin in the European Union, Norway, Iceland and Liechtenstein – 2001 Report. DG SANCO, Food and Veterinary Office, Brussels, Belgium.

Food & Drug Administration (1996). *E. coli* 0157:H7 *Outbreak Associated with Odwalla Brand Apple Juice Products.* FDA, Department of Health and Human Services, Washington, DC.

Gunning, A.P., Bongaerts, R.J.M. and Morris, V.J. (2009). Recognition of galactan components of pectin by galectin-3. *FASEB Journal* **23**, 415–424.

Hargreaves, J.C., Adl, M.S., Warman, P.R. and Rupasinghe, H.P.V. (2008). The effects of organic and conventional nutrient amendments on strawberry cultivation: fruit yield and quality. *Journal of the Science of Food and Agriculture* **88**, 2669–2675.

Hecke, K., Herbinger, K., Veberic, R., *et al.* (2006). Sugar-, acid- and phenol contents in apple cultivars from organic and integrated fruit cultivation. *European Journal of Clinical Nutrition* **60**, 1136–1140.

Irish National Food Residue Database (2009). Residue Studies Group, Food Safety Department, Ashtown Food Research Centre, Teagasc, Ireland. http://nfrd.teagasc.ie, accessed October 11, 2010.

Jackson, C.L., Dreaden, T.M., Theobold, L.K., *et al.* (2007). Pectin induces apoptosis in human prostate cancer cells: correlation of apoptotic function with pectin structure. *Glycobiology* **17**, 805–819.

Jorhem, L. and Slanina, P. (2000). Does organic farming reduce the content of cadmium and certain other trace metals in plant foods? A pilot study. *Journal of the Science of Food and Agriculture* **80**, 43–48.

Karavoltsos, S., Sakellari, A., Dimopoulos, M., *et al.* (2002). Cadmium content in foodstuffs from the Greek market. *Food Additives and Contaminants* **19**, 954–962.

Karavoltsos, S., Sakellari, A., Dassenakis, M. and Scoullos, M. (2008). Cadmium and lead in organically produced foodstuffs from the Greek market. *Food Chemistry* **106**, 843–851.

Kihlberg, I. and Risvik, E. (2007). Consumers of organic foods – value segments and liking of bread. *Food Quality and Preference* **18**, 471–481.

Kirchmann, H. and Thorvaldsson, G. (2000). Challenging targets for future agriculture. *European Journal of Agronomy* **12**, 145–161.

Kumpulainen, J. (2001). Organic and conventional grown foodstuffs: nutritional and toxicological quality comparisons. *Proceedings of the International Fertiliser Society* **472**, 1–20.

Lecerf, J.M. (1995). Biological agriculture: interest for human nutrition? *Cahiers de Nutrition et de Dietetique* **30**, 349–357.

Lester, G.E., Manthey, J.A. and Buslig, B.S. (2007). Organic vs conventionally grown Rio Red whole grapefruit and juice: comparison of production inputs, market quality, consumer acceptance, and human health-bioactive compounds. *Journal of Agricultural and Food Chemistry* **55**, 4474–4480.

Lindén, A., Andersson, K. and Oskarsson, A. (2001). Cadmium in organic and conventional pig production. *Archives of Environmental Contamination and Toxicology* **40**, 425–431.

Lombardi-Boccia, G., Lucarini, M., Lanzi, S., *et al.* (2004). Nutrients and antioxidant molecules in yellow plums (*Prunus domestica* L.) from conventional and organic productions: a comparative study. *Journal of Agricultural and Food Chemistry* **52**, 90–94.

Magkos, F., Arvaniti, F., and Zampelas, A. (2003). Putting the safety of organic food into perspective. *Nutrition Research Reviews* **16**, 211–221.

Malmauret, L., Parent-Massin, D., Hardy, J.L. and Verger, P. (2002). Contaminants in organic and conventional foodstuffs in France. *Food Additives and Contaminants* **19**, 524–532.

Moolenaar, S.W. (1999). Heavy-metal balances, Part II. Management of cadmium, copper, lead, and zinc in European agro-ecosystems. *Journal of Industrial Ecology* **3**, 41–53.

Moolenaar, S.W. and Lexmond, T.M. (1999). Heavy-metal balances, Part I: General aspects of cadmium, copper, zinc and lead balance studies in agro-ecosystems. *Journal of Industrial Ecology* **2**, 45–60.

New Zealand Food Safety Authority (2009). 2009 New Zealand Total Diet Study. http://www.nzfsa.govt.nz/publications/media-releases/2009/2009-09-22-tds-second-quarter-results.htm, accessed March 2, 2010.

Ninfali, P., Bacchiocca, M., Biagiotti, E., *et al.* (2008). A 3-year study on quality, nutritional and organoleptic evaluation of organic and conventional extra-virgin olive oils. *Journal of the American Oil Chemists' Society* **85**, 151–158.

Piemontese, L., Solfrizzo, M. and Visconti, A. (2005). Occurrence of patulin in conventional and organic fruit products in Italy and subsequent exposure assessment. *Food Additives and Contaminants* **22**, 437–442.

Reganold, J.P., Glover, J.D., Preston, P.K. and Hinman, H.R. (2001). Sustainability of three apple production systems. *Nature* **410**, 926–930.

Reig, G., Larrigaudière, C. and Soria, Y. (2007). Effects of organic and conventional growth management on apple fruit quality at harvest. *Acta Horticulturae* **737**, 61–65.

Róth, E., Berna, A., Beullens, K., *et al.* (2007). Postharvest quality of integrated and organically produced apple fruit. *Postharvest Biology and Technology* **45**, 11–19.

Roussos, P.A. and Gasparatos, D. (2009). Apple tree growth and overall fruit quality under organic and conventional orchard management. *Scientia Horticulturae* **123**, 247–252.

Siderer, Y., Maquet, A. and Anklam, E. (2005). Need for research to support consumer confidence in the growing organic food market. *Trends in Food Science and Technology* **16**, 332–343.

Tarozzi, A., Morroni, F., Cantelli-Forti, G., *et al.* (2006). Antioxidant effectiveness of organically and non-organically grown red oranges in cell culture systems. *European Journal of Nutrition* **45**, 152–158.

Trewavas, A. (2001). Urban myths of organic farming. *Nature* **410**, 409–410.

Tsatsakis, A.M., Tsakiris, I.N., Tzatzarakis, M.N., *et al.* (2003). Three-year study of fenthion and dimethoate pesticides in olive oil from organic and conventional cultivation. *Food Additives and Contaminants Part A* **20**, 553–559.

UK Pesticide Residues Committee (2010). Report for 2009. http://www.pesticides.gov.uk/prc_home.asp, accessed July 4, 2010.

USDA (2007). US Department of Agriculture Pesticide Data Program, annual summary, calendar year 2006. USDA, Washington, DC. http://www.ams.usda.gov/pdp, accessed July 7, 2010.

USDA (2008). US Department of Agriculture Pesticide Data Program, annual summary, calendar year 2007. USDA, Washington, DC. http://www.ams.usda.gov/pdp, accessed July 7, 2010.

USDA (2010). US Department of Agriculture Pesticide Data Program. http://www.ams.usda.gov/pdp, accessed April 26, 2010.

Veberic, R., Trobec, M., Herbinger, K., *et al.* (2005). Phenolic compounds in some apple (*Malus domestica* Bork) cultivars of organic and integrated production. *Journal of the Science of Food and Agriculture* **85**, 1687–1694.

Vian, M.A., Tomao, V., Coulomb, P.O., *et al.* (2006). Comparison of the anthocyanin composition during ripening of Syrah grapes grown using organic or conventional agricultural practices. *Journal of Agricultural and Food Chemistry* **54**, 5230–5235.

Wang, S.Y., Chen, C.T., Sciarappa, W., *et al.* (2008). Fruit quality, antioxidant capacity, and flavonoid content of organically and conventionally grown blueberries. *Journal of Agricultural and Food Chemistry* **56**, 5788–5794.

Woese, K., Lange, D., Boess, C. and Bogl, K.W. (1997). A comparison of organically and conventionally grown foods – results of a review of the relevant literature. *Journal of the Science of Food and Agriculture* **74**, 281–293.

Worthington, V. (2001). Nutritional quality of organic versus conventional fruits, vegetables and grains. *Journal of Alternative and Complementary Medicine*, **7**, 161–173.

Yildirim, H.K., Akcay, Y.D., Guvenc, U., *et al.* (2005). Antioxidant activities of organic grape, pomace, juice, must, wine and their correlation with phenolic content. *International Journal of Food Science and Technology* **40**, 133–142.

5 Cereal Grains

Consumers have concerns about the possibility of chemical and pesticide residues in cereal grains, causing them often to seek out organic grains. This is understandable since the organic regulations restrict the types of pesticides that can be used, presumably leaving fewer residues in the grains. Reassuringly, though, this is less of a concern once the relevant facts are established.

Pesticide residues

A main concern is the possibility of pesticide residues in grains, either as a result of crop spraying and dusting, residues in the soil, or residues of pesticides used to protect harvested grain during storage.

The US Department of Agriculture tested a total of 5512 samples of domestically produced food (Table 5.1) and imported food (Table 5.2) from 85 countries for pesticide residues in 2006. Among grains and grain products, the violation rate was zero for domestic samples and 2.2 percent for imports. These findings indicate that pesticide residue levels in grain foods are generally well below EPA (Environmental Protection Agency) tolerances, corroborating earlier findings. Imported rice, however, was more of a concern, with 12 percent of samples being in violation of the EPA standards.

According to the latest data from this agency (USDA, 2010), 650 samples of corn grain (domestic) and 184 samples of rice were tested in 2008. The percentage of total residue detections for both corn grain and rice was 0.7 percent, with no samples in violation. Fifteen different residues were detected in the corn grain samples. The most frequently detected residue was malathion, which was detected in 219 samples (33.7 percent). Chlorpyrifos was detected in 116 samples (17.8 percent) and other compounds were detected in more than 1 percent of the samples. No residue exceeded the established tolerance. No comparable values for organic grains were available.

Thirteen different residues were detected in the rice samples. The most frequently detected residue was piperonyl butoxide, which was detected in 57 samples (31.0 percent). MGK-264 (N-octyl bicycloheptene dicarboximide)

Organic Production and Food Quality, First Edition. Robert Blair.
© 2012 John Wiley & Sons, Ltd. Published 2012 by John Wiley & Sons, Ltd.

Table 5.1   Frequency of pesticide residues in US domestic grain and grain products, excluding the residues of banned organochlorines (USDA, 2007: http://www.ams.usda.gov/pdp)

Product	Total samples analyzed	Percent free of residues	Samples in violation	Violations over tolerance	Violations no tolerance
Corn (maize) and corn products	52	94.2	0	0	0
Oats and oat products	5	100	0	0	0
Rice and rice products	13	69.2	0	0	0
Soybeans and soybean products	14	100	0	0	0
Wheat and wheat products	62	79.0	0	0	0
Other grains and grain products	4	100	0	0	0
Breakfast cereals	2	100	0	0	0
Bakery products, crackers, etc.	6	83.3	0	0	0
Snack foods	2	100	0	0	0
Total	160	86.9	0	0	0

Table 5.2   Frequency of pesticide residues in US imported grain and grain products, excluding the residues of banned organochlorines (USDA, 2007: http://www.ams.usda.gov/pdp)

Product	Total samples analyzed	Percent free of residues	Samples in violation	Violations over tolerance	Violations no tolerance
Barley and barley products	9	100	0	0	0
Corn (maize) and corn products	6	100	0	0	0
Oats and oat products	12	100	0	0	0
Rice and rice products	25	72.0	12.0	0	3
Wheat and wheat products	21	90.5	0	0	0
Other grains and grain products	21	76.2	0	0	0
Breakfast cereals	8	75.0	0	0	0
Bakery products, crackers, etc.	23	82.6	0	0	0
Pasta and noodles	6	100	0	0	0
Snack foods	7	100	0	0	0
Total	138	85.5	2.2	0	3

Table 5.3   Results of Australian Food Monitoring Program for chemical and pesticide residues in cereal grains (Department of Agriculture, Food and Fisheries, 2004–2005)

Grain product	Total samples	Total analyses	Residues exceeding standard	Residues found	Environmental contaminants exceeding standard
Cereal grains	4386	116 623	16	0	1
Flour and bran	188	5 108	1	0	0
Pulses (peas, beans)	166	4 761	1	0	0
Canola	234	6 365	1	0	0

(*Source*: Australian Department of Agriculture, Fisheries and Forestry (2010). Plant Products: Random Residue Monitoring results. http://www.daff.gov.au/agriculture-food/nrs/programme. Reprinted with permission of the Department of Agriculture, Fisheries and Forestry.)

was detected in 27 samples (14.7 percent). Other compounds were detected in 1 percent or more of the samples. As with corn grain, no residue exceeded the established tolerance. No comparable values for organic grains were available from this monitoring study.

Australia has a very good food monitoring program. Data from the 2004–2005 testing period are shown in Table 5.3. The grain monitoring project covered 10 commodities, with 4974 samples being collected and 132 857 analyses undertaken. Wheat grain and its products (bran and flour) made up the largest number of samples. The other grain products included barley, sorghum, lupin seed, field peas, oats and chickpeas.

Only 17 samples were found not to comply with Australian Food Standards. There were 20 residues of agricultural chemicals detected above the Australian Standards for this class of foodstuffs. No residue violations were found in any samples of barley, chickpeas, field peas or lupinseed.

Tests for a range of environmental metal contaminants were conducted on 365 samples. One sample of barley was found to contain a lead level (0.26 ppm) higher than the Australian Standard (0.2 ppm). However, further sampling of the grain showed the lead level to be below the Australian Standard.

Residue monitoring in 2008–2009 for the grains program covered 21 cereal, pulse and oilseed commodities. Wheat (55 percent) and barley (18 percent) comprised the largest proportion of samples collected. The total number of cereal grains analyzed was 3795, with 99.5 percent meeting Australian standards.

The Danish National Pesticide Survey on grain for human consumption tested for residues of two plant growth regulators used to stabilize stalks in cereals (Juhler and Vahl, 1999). Of 77 samples analyzed, 51 contained residues of one chemical and 11 contained residues of the other. However, the residues were in all cases below the maximum residue limits. The highest levels were found in oatmeal (3.76 ppm) and rye (1.08 ppm).

Similar results were reported from Germany (Gans *et al.*, 2000) where cereal and oilseed crops were treated with growth regulator chemicals at various stages during growth. Residues were found in the grains, but at levels well under the maximum allowed.

All of the above data suggest that grains do contain residues, but that the residues are too small to be of importance in relation to human health. Many of the residues found were at the limit of detection. More studies need to be carried out to provide comparable data on organic and regular grains. Some consumers prefer to purchase organic grain rather than regular grain, but that decision has to be viewed as being based on preference rather than proven facts about residues.

Countries need to start testing organic foods separately in their food monitoring programs in order to clarify this issue.

Chemical residues

A related concern about cereal grains is that growing cereals might take up chemicals in the environment and deposit them in the harvested grain. These chemicals include some metals and some persistent pesticides remaining in the environment as a result of past use.

Organic soils should be lower in chemical residues because of the use of manure as fertilizer and the avoidance of most pesticides. However, it is difficult to assess whether this results in any difference in the amount of residues in the grain, since comparable data are lacking.

The chemicals of most concern are the heavy metals. The term "heavy metal" refers to any metallic chemical element that has a relatively high density and is toxic or poisonous at low concentrations. Heavy metals are natural components of the Earth's crust. Examples are copper, selenium and zinc, which are dietary requirements in trace amounts for humans and animals but are toxic at high levels. Most regions of the world, for instance, have areas where soil selenium levels are high. Plants growing on these seleniferous soils absorb excessive amounts of selenium, leading to toxicity in animals that consume them.

Mercury, cadmium, arsenic, chromium, thallium, and lead are other heavy metals, which are not required in the diet and are toxic. Lead, cadmium, and mercury are regarded as environmental pollutants and considered dangerous because they tend to accumulate in the tissues of humans, animals and plants.

Human exposure to the polluting heavy metals is mainly from the drinking water and air, pollution from mining and smelting operations, the burning of fossil fuels, cigarette and cannabis smoke, municipal wastes, fertilizers, pesticides and sewage. Consequently efforts are now made to minimize the risks in the environment from toxic heavy metals.

Food is a minor source of these elements, except seafood (e.g. shellfish), which is known to accumulate heavy metals from the marine environment.

Not all people fear chemicals in the environment. Some people in fact seek them out, a practice that has been continued for centuries. The King's Bath, in Bath, England, which was first developed by the Romans, still delivers about a quarter of a million liters of water a day at a temperature of 46.5 °C. The water in the King's Bath contains some thirty minerals, including calcium, bismuth, magnesium, iron, lead, potassium, strontium and sulfur. In addition, it is slightly radioactive. Traditionally immersion was prescribed for rheumatic and urinary diseases, and drinking was prescribed for internal ailments.

Cadmium has become a more prevalent cause for concern in recent years. Like lead, it is an underground mineral that did not enter our air, food, and water in significant amounts until it was mined as part of zinc deposits. Now there is some concern about environmental contamination with this element. Cadmium and zinc are found together in natural deposits and are similar in structure and function in the body. Cadmium may displace zinc in some of its important enzymatic and organ functions, and thereby interferes with them. The zinc:cadmium ratio in the diet is important, cadmium toxicity being greatly increased with zinc deficiency. Adequate levels of zinc protect against tissue damage by cadmium.

Research was conducted on the cadmium contents of oats over a 3-year period in Finland by Eurola *et al.* (2003). No significant differences were found between samples grown organically or conventionally. Large seasonal and regional variations were found in the cadmium concentrations. In official variety trials, the average cadmium contents in 1997, 1998 and 1999 were 0.046, 0.029 and 0.052 mg/kg dry weight, respectively, the range being 0.008 to 0.120 mg/kg. The concentrations were generally well below the maximum permitted level of 0.100 mg/kg fresh weight. Nitrogen fertilization increased the cadmium contents of oats. Some varieties had consistently higher contents of this element, prompting the researchers to suggest that it was possible to cultivate and develop oat varieties less likely to accumulate cadmium.

Other issues relating to grains

An important issue for consumers to consider is mycotoxin contamination, which in organic and regular grains has caused serious problems worldwide. Mycotoxins are toxins produced by fungi (molds). For instance, Biffi *et al.* (2004) found ochratoxin in all 211 cereal products tested (flours and bakery products), but mostly at very low levels and with no major differences due to cereal production system (conventional, organic or integrated pest management).

Mycotoxins: are organic grains less safe?

As with other foods, the possibility that organic grains may be less safe than regular foods seems a strange question to consider. However, organic grain crops may suffer mold damage because they have not been protected by spraying. The mold infestation may occur in the field or during storage of the crop and can result in the development of mycotoxins in the grain. Some mycotoxins are poisonous to humans and farm stock.

An Italian study found that the occurrence of deoxynivalenol contamination was higher than 80 percent in both organic and conventional foods (Cirillo *et al.*, 2003). Fumonisin B(1) was found in 20 percent of organic foods and in 31 percent of conventional foods, and fumonisin B(2) in more than 32 percent of the food samples from both types. The highest median concentration of deoxynivalenol occurred in conventional rice-based foodstuffs (207 µg/kg), fumonisin B(1) in conventional corn-based foods (345 µg/kg) and fumonisin B(2) in organic wheat-based foods (210 µg/kg).

A total of 205 cornflake samples collected in Belgian retail stores during 2003–2004 were surveyed by Paepens *et al.* (2005) for the natural occurrence of fumonisin B1 (FB1), B2 (FB2), and B3 (FB3). They were from both conventional and organic production. FB1 concentrations ranged from not detected to 464 µg/kg with a mean concentration of 104 µg/kg. For FB2 and FB3, the concentration ranges varied from not detected to 43 µg/kg and from not detected to 90 µg/kg, respectively. The mean concentrations for FB2 and FB3 were 12 and 21 µg/kg, respectively. Based on the findings it was concluded that the production system (organic vs. conventional) did not have a significant effect on the fumonisin concentrations but that the variation between different batches was highly significant.

A recent Dutch study compared contaminants and microorganisms in Dutch organic and conventional foods (Hoogenboom *et al.*, 2008). Organic products were analyzed for the presence of contaminants, microorganisms and antibiotic resistance and compared with those from conventional products. No differences were observed in the content of the fusarium toxins deoxynivalenol and zearalenone between organic and conventional wheats, during both a dry period and a very wet period which promoted the production of these toxins.

Analysis of data from the 2002–2005 harvests found a lower concentration of some fusarium toxins in organic oats (Edwards, 2009). Ten trichothecenes were measured: deoxynivalenol (DON), nivalenol, 3-acetylDON, 15-acetylDON, fusarenone X, T-2 toxin (T2), HT-2 toxin (HT2), diacetoxyscirpenol, neosolaniol and T-2 triol. Three (15-acetylDON, fusarenone X and diacetoxyscirpenol) were not detected, and moniliformin and zearalenone were absent or rarely detected. HT2 and T2 were the most frequently detected fusarium mycotoxins and were found to be present above the limit of quantification ($10\,\mu g/kg$) in 92 and 84 percent of samples, respectively. These mycotoxins were usually present at the highest concentrations. Year and region had a significant effect on HT2 + T2 concentration. More importantly the average concentration of this class of mycotoxin was found to be five times lower in organic samples than in conventional samples. A tentative upper limit of $500\,\mu g$ HT2 + T2 per kg is under consideration for oats intended for human consumption in Europe. According to the findings presented by Edwards (2009), 36 percent of conventional samples and 11 percent of organic samples of oats exceeded this limit. No samples exceeded the legal limits for DON or zearalenone. The report has disturbing implications for the oats industry in the UK.

Probably the best-known mycotoxin is ergot, because of its hallucinogenic properties and its association with witches. This species of mold was also the original source from which LSD was first isolated. It is believed that symptoms of ergotism have been recorded since the Middle Ages and possibly even as far back as ancient Greece. The mold is mostly associated with rye but can affect other grains. In high doses it can be lethal. Ergotism is still a disease of public health importance, particularly in developing countries. One theory is that the ergot problem in rye, which can be seen because of the presence of blackish 'bunts' on the grain, led to the rise in popularity of white bread in Western countries.

There do not appear to be documented scientific reports of ergot in organic grains, but one man recounted his unfortunate experience with organic rye on July 31, 2008, at a supermarket in Newport Beach, California, where he was arrested (Zenor, 2008). The story is all the more unusual because this man is a trained chemist. The name of the store where he purchased the organic rye has been omitted to avoid embarrassment.

> I have always been interested in wild foods and mushrooms were on top of my list of favorite wild foods. I recently started a hobby of growing mushrooms. After some research, I engineered a small scale operation to extract sterile samples of mushroom tissue and grow it on potato dextrose agar which was easy and cheap to make. I found the organic rye grain . . . in Huntington Beach. The rye grain is used to increase the amount of mycelium or fungal cells. The grain can easily be prepared to create an ideal environment for the cultivation of the fungal cells. Rye is the best grain apparently because it doesn't burst as easily or something like that . . . My agar inoculated with mycelium was reaching a point where it could be expanded by adding it to prepared grain so I gathered up the rye and winter wheat to see if I could prepare some sterilized grain to add mycelium to and make some grain spawn. I still had most of the rye and a few pounds of wheat so I decided to throw it all in together and start to soak it. When I added the rye, I noticed that it contained several dark kernels. They were fairly easy to see in the even sized kernels of grain. I used to eat

wild oats on long walks and I was aware of fungus called ergot that also sometimes grows on wild oats in South Dakota. I have never seen it on the local oats. I remember tasting the ergot as a child because it looked like a deformed grain of oats and I ate a few of them without noticing any effect. That is why I wasn't particularly worried about tasting the kernels . . . I was mainly worried about some strange fungus getting in my grain and doing something to spoil it so I carefully went through the grain and removed all the ergot and threw them away. I tasted a few of them because I was trying to figure out if it was actually ergot. It had exactly the same texture and taste so I was absolutely sure it was ergot . . . A crude example of what it is like is if you took a small bit of pancake batter and placed it on a hot griddle for about ½ hour.

I didn't know what caused my bizarre behavior and didn't even think of the possibility of ergot poisoning until I overheard a conversation in the room about "magic mushrooms" a few days later. That made me think about mind altering effects of ergot. I realized that it was only a couple of hours or less after I purposefully tasted the ergot and ate some of the moist grain before I went to [store name]. Soaking in water could have allowed some of the ergot to increase in concentration or perhaps it was contaminated. After the fact, it seems like a very dangerous thing to do since molds and fungus grow so readily on moist rye grain . . . .

The reason I went all the way to [store name] was in the hope that they would have some good samples of either oyster mushrooms or shiitake to add some new cultures in case the ones I had went bad. [Store name] at Newport Coast was the only place I have found good fresh oyster mushrooms and I didn't have a clean batch of mycelium of that species. That was why I went there but I now believe I was severely affected by ergot poisoning so I don't really remember exactly what was happening or how it happened. I was most definitely then under the influence of some very powerful mind altering substance and my mind failed to inform me of that fact. I didn't have my normal capacity to realize something was wrong.

I remember how strange it felt going through the store but there was no particular useful or sane purpose in anything I did. Tequila is something I consider repulsive to drink and I can't even imagine actually wanting to drink it. I remember running and hearing voices of authority behind me. I also remember hoping they would go away if I put down what I was carrying. They didn't go away and some wording of "Stop, stop," clearly registered in my mind. I tried to obey and all I remember after that is the world spinning forward out of control and a very powerful impact with the ground that made me senseless. I tried to get up but was not able to more than stumble a few steps as near as I remember . . .

I was still significantly impaired 4 days later where I just couldn't remember my cell block number. I carried it on a piece of paper but it was thrown away during a search. I was worried that I would get into trouble because I didn't know where I belonged. The cell block was A1 if I can trust my memory. Fortunately, it didn't require a lot of thought being there and my cellmates took care of me and always had me follow someone. It seemed natural since I was new but remembering now, it sounds like I was almost retarded. I am actually worrying about permanent brain damage but I think most likely the long term memory loss will be restricted to the unpleasant aspects of incarceration. I am always the optimist.

So, beware of tainted rye! Luckily the molded grains can be spotted easily. Organic growers and merchants need to be aware of this hazard and ensure that they have an adequate quality-control program in use to avoid mycotoxin poisoning by consumers.

Mycotoxin contamination of foods is common worldwide, especially in warm, humid regions. For instance Morocco, which is a North African country surrounded by the Mediterranean Sea and Atlantic Ocean, has a climate characterized by high humidity and high temperature which favors growth of molds. Studies on foods produced there (cereals, bread, milk, spices, wine, olives, poultry feeds, dried fruits and nuts) have shown an average contamination rate often above 50 percent.

Parts of North America and Europe also have conditions that favor the contamination of crops with mycotoxins.

Contamination of organic foods with fumonisin, another of the mycotoxins, has been reported although the effects are not as dramatic as those described above for ergot. This mycotoxin has been linked to esophageal cancer in humans in parts of South Africa and China where consumption of fumonisin-contaminated corn (maize) is common. In addition to damaging the brain, liver and lungs, fumonisins also affect the kidneys, pancreas, testes, thymus, gastrointestinal tract, and blood cells. There is also concern that consumption of fumonisins during early pregnancy may result in an elevated risk of neural-tube defect in the developing fetus. This was the suspected cause of problems reported recently in babies born to mothers in the Rio Grande Valley in southern Texas who ate large amounts of corn tortillas during pregnancy. The FDA limit on fumonisins in human food is 2–4 ppm and in animal feed it is 5–100 ppm.

Published reports on fumonisin contamination indicate no clear differences between organic and conventional corn flakes or other organic cereal products. A research team from Finland and Italy (Jestoi *et al.*, 2004) tested the contamination levels of 16 different mycotoxins in randomly selected organic and conventional grain-based products purchased from Finnish and Italian markets. Overall the concentrations of mycotoxins were low in all of the samples. Neither geographical origin nor type (organic or conventional) had any influence on mycotoxin concentration. Baby foods were found to have significantly lower concentrations of mycotoxins than the other products, with an average total mycotoxin content of 47 µg/kg compared with 99 µg/kg for the other kinds of food. This last finding suggests that higher quality cereal products go into baby foods.

Studies on other mycotoxins indicate similar mixed findings. Some mycotoxins show up in beer, which is logical since beer is made from grains. The concentration of zearalenone was found to be similar in organic and conventional cereal products in one report. Other studies found higher levels in conventional wheat and cereal products, and a higher level in organic beer in one year but not the next (Anselme et al., 2006). Higher levels of zearalenone were reported in conventional wheat than in organic wheat. Higher levels of ochratoxin were reported in organic wheat and in organic beer in one year but not the next, and no differences were found between organic and conventional cereal products. Higher levels of aflatoxin were found in some, but not all, samples of organic milk in Italy.

The overall conclusion that can be drawn from these results is that there are no consistent differences between organic and conventional grains in mycotoxin content. It is likely that growing and harvesting conditions have a much greater influence on the likelihood of mycotoxin contamination than the production system (organic vs. conventional). The findings also suggest that farmers need to be vigilant in ensuring that mycotoxin risks to the consumer are minimized. Recommendations include selecting crops that are adapted to local climatic and soil conditions and making sure that crops are not stressed during pre-harvest to the extent that they become infected by molds. Insect damage to crops can also lead to mold infestation, therefore insect control is necessary. Quality control should be exercised to reject foods that might be contaminated with mycotoxins.

Consumers can minimize the risk of mycotoxin contamination in crops such as corn (maize) by choosing ears that are free of mold and are fresh rather than from storage.

Gene-modified crops

The use of biotechnology to improve genetic makeup has become widely adopted in breeding new and improved crops such as corn (maize). For instance Brookes (2008) reviewed the impact of using gene-modified (GM) insect-resistant maize in Europe since 1998. In 2006 there were plantings of this crop in seven EU member states. An economic analysis showed an improvement in profitability of between 12 and 21 percent. Insecticide spraying had also been reduced and the grain quality was improved as a result of lower mycotoxin contamination.

However, some consumers worry that the cereal grains they buy may have come from GM crops, preferring food from traditional crops unaltered by biotechnology. Organic producers have a declared objection to this technology, and consumers who have a similar objection therefore opt for organic grains. Objection to GM grains is especially strong in Europe where the majority of consumers prefer heritage or traditional varieties of grains.

Not all societies that are considered very traditional reject crops derived by biotechnology, however. For example, the Muslim Council of Indonesia (Indonesian Ulemas Council) approved GM foods in 2003. Also, GM corn is being grown by Amish farmers in the USA, a group that follows a very traditional lifestyle.

The public attitude to crop variety improvement by GM methods may be dependent, therefore, on the type and accuracy of the information available to them.

An example of the dilemma facing farmers over GM crops is the potato blight experienced in the UK in 2007. Organic crops were sprayed with copper compounds to treat the fungal infection, but this treatment was decried by other farmers who claimed that copper contamination was a greater damage to the environment than the use of GM varieties resistant to the blight. The potato blight problem referred to above brings to mind the most famous loss of an organic potato crop due to fungal disease: the potato failure of 1845 that resulted in widespread starvation in Ireland and a loss of 20–25 percent of the people.

What emerges from a consideration of the facts surrounding GM foods is that support or opposition to GM technology is not based simply on established scientific facts. The situation is complicated by biases, ethical and moral considerations and international trade considerations.

Consumers who wish to avoid GM foods on principle will therefore opt to buy organic foods or meat, milk and eggs from animals that have not been fed GM feeds. Consumers who cannot afford the more expensive organic foods or who are willing to adopt the benefits of biotechnology will opt for conventional foods without compromising their health.

Nutritional and organoleptic qualities

Much less information has been published on the nutritional quality of cereal grains produced organically than on vegetable produce or fruit. This is in spite of the fact that grains are produced for the important animal feed market as well as for human consumption.

Hornick (1992) reviewed literature findings on factors affecting the nutritional quality of crops and concluded that earlier research to investigate the nutritional status of crops grown with either chemical fertilizers or organic fertilizers often gave contradictory results on crop yields and on the content of mineral and vitamin contents. She outlined several research needs to ensure that comparable results and valid comparisons can be obtained. Woese *et al.* (1997) conducted a subsequent analysis of the published data and concluded that there was a trend to lower protein contents in the organically produced cereals, mainly wheat and rye. Overall quality was also reported to be lower in organic grains.

Wheat

Starling and Richards (1990) reported the results of a cereal quality survey conducted by the UK Home Grown Cereal Authority (HGCA). It was found that organic wheat tended to have a higher specific weight (weight of a fixed mass of grain, e.g. bushel weight: a low weight indicates lower quality), similar Hagberg falling number and lower protein content than conventional wheat. The same research group (Starling and Richards, 1993) reported on the quality of commercial samples of organically grown wheat harvested in 1988–1992. Spring wheat cv. Axona and winter wheat cv. Maris Widgeon were most successful in achieving bread-making quality, although Maris Widgeon had low Hagberg falling number in poor harvest years. Protein content was lower in organic than in conventionally grown wheat.

An Australian study (Ryan *et al.*, 2004) compared the yield and grain mineral concentrations of wheat grown under organic and conventional management. The yield of conventional grain was found to be 17–84 percent higher. Minor variations occurred in the nitrogen, potassium, magnesium, calcium, sulfur and iron concentrations of the two wheats. Conventional grain had lower zinc and copper contents, but higher manganese and phosphorus contents than organic grain.

A greater amount of data was obtained by Mäder *et al.* (2007), who conducted a 21-year study in Europe to test the effects of organic and conventional farming practices on wheat quality. Samples were taken from three of the harvests and tested for nutritional content, milling and baking quality and acceptability in food preference studies with rats fed biscuits made from the wheats. Results are shown in Table 5.4.

Yield was found to be 14 percent lower with organic production but nutritional value (protein content, amino acid composition and mineral and trace element contents) and baking quality were not affected consistently by farming system. Incidence of mycotoxin contamination was low and did not differ in the two different types of wheat. In food preference tests, rats preferred organically over conventionally produced wheat. Tests involved a comparison of wheat grown organically or grown conventionally but with manure used in fertilizing the fields. No data were provided on a comparison of the acceptability of wheats from organic and conventional production, i.e. mineral fertilizer only being in the field managed conventionally.

Table 5.4 Effects of organic production on wheat quality (from Mäder *et al.*, 2007; © John Wiley & Sons)

Parameter	Organic	Conventional	Significance of difference
Yield (tonnes per hectare, DM basis)			
Harvest 1	3.57	3.61	NS
Harvest 2	4.06	5.15	$P < 0.001$
Harvest 3	4.11	4.77	$P < 0.001$
Protein yield (kg/ha)			
Harvest 1	512	495	$P < 0.01$
Harvest 2	485	657	$P < 0.001$
Harvest 3	520	686	$P < 0.001$
Grain composition (g/kg DM unless stated)			
Crude protein			
Harvest 1	143.7	138.9	$P < 0.05$
Harvest 2	124	131.4	NS
Harvest 3	126.4	145.1	$P < 0.001$
Amino acids in harvest 1 (g/kg total protein)			
Lysine	25.8	25.8	NS
Methionine	12.9	13.5	NS
Threonine	31.1	30.2	NS
Tryptophan	12.0	11.7	NS
Ash (g/kg DM)			
Harvest 1	19.9	21.3	NS
Harvest 2	18.5	17.6	NS
Harvest 3	16.7	16.7	NS
P (g/kg DM)			
Harvest 1	4.30	4.28	NS
Harvest 2	4.08	4.03	NS
Harvest 3	3.73	3.65	NS
K (g/kg DM)			
Harvest 1	4.46	4.65	NS
Harvest 2	4.80	4.78	NS
Harvest 3	4.53	4.38	NS
Ca (g/kg DM)			
Harvest 1	0.50	0.52	NS
Harvest 2	0.45	0.44	NS
Harvest 3	0.44	0.48	NS
Mg (g/kg DM)			
Harvest 1	1.28	1.30	NS
Harvest 2	1.26	1.19	NS
Harvest 3	1.24	1.19	NS

(*Continued*)

Table 5.4   (*Continued*)

Parameter	Organic	Conventional	Significance of difference
		Mn (mg/kg DM protein)	
Harvest 1	35.3	35.9	P < 0.05
Harvest 2	37.4	42.5	P < 0.001
		Zn (mg/kg DM protein)	
Harvest 1	36.9	35.9	NS
Harvest 2	33.7	32.2	NS
		Cu (mg/kg DM protein)	
Harvest 1	4.71	5.49	P < 0.01
Harvest 2	6.52	6.45	NS
		Mo (mg/kg DM protein)	
Harvest 1	0.26	0.25	NS
		Co (mg/kg DM protein)	
Harvest 1	0.18	0.19	NS

NS = Not significant at P<0.05.

In a subsequent study Zörb *et al.* (2009) grew wheat organically and conventionally and measured responses in grain size and composition. Total yield was not measured. Results are shown in Table 5.5.

It was found that 1000-seed weight and crude protein content were higher in conventional wheat. Differences in mineral content were not consistent. Some sugars were higher in the conventional wheat, and no difference was observed in antioxidant content between wheat types.

Table 5.5   Effects of organic and conventional production methods on the composition of wheat grain (from Zörb *et al.*, 2009; © American Chemical Society)

Component (in dry matter)	Organic	Conventional	Significance of difference
1000-seed wt (g)	40	47	P < 0.05
Protein (percent)	10.7	14.1	P < 0.05
K (mg/kg)	4557	3972	P < 0.05
Ca (mg/kg)	270	295	NS
Mg (mg/kg)	1215	1200	P < 0.05
Na (mg/kg)	62	61	NS
Sucrose (mg/kg)	5.83	5.40	NS
Maltose (mg/kg)	1.56	1.26	P < 0.05
Fructose (mg/kg)	0.27	0.36	P < 0.05
Glucose (mg/kg)	0.15	0.19	P < 0.05
Anti-oxidative capacity (µmol/g Trolox equivalents)	14	13.6	NS

NS = Not significant at P<0.05.

Components other than main nutrients and which have been linked to human health have been studied. Zuchowski *et al.* (2009) measured the content of phenolic acids in four winter wheat cultivars grown using conventional and organic agricultural practices. Five phenolic acids were detected, with ferulic acid being the predominant phenolic acid in all samples. The remaining phenolic acids, i.e. sinapic acid, *p*-coumaric acid, vanillic acid, and *p*-hydroxybenzoic acid, were present in considerably lower amounts. Significant differences among cultivars in the concentration of particular phenolic acids, as well as in the total phenolic acid content, were observed. The effect of various agricultural practices on phenolic acid levels in wheat grains was also analyzed. Organic production was found to result in a small, statistically insignificant trend towards a higher level of phenolic acids.

Some information has been obtained on the baking quality of organic wheat. Haglund *et al.* (1998) conducted an evaluation of wholemeal bread from organic and conventionally grown wheat. The purpose of the project was to study how conventional and ecological farming systems and different dough kneading intensity affected the baking properties of wholemeal flour, and how those properties affected the taste and consistency of wholemeal bread. Sensory evaluations were performed with respect to wholemeal loaves from winter wheat. The dough from each wheat sample was divided into two parts. One part was subjected to low kneading intensity, the other to high kneading intensity. High kneading intensity refers to standard commercial practices. Wholemeal from the conventional farming system had a higher protein content than wholemeal from ecological farming systems. Wholemeal from the conventional farming system resulted in bread with a large volume and a high degree of elasticity while wholemeal from ecological farming systems resulted in a dry bread. High kneading intensity generally resulted in a dry and less elastic bread which had a significantly stronger tinge of grey on the surface of the slice.

A novel approach to the organoleptic quality of organic wheat was adopted in a study in the UK. In a 3-year study, researchers tested the preference of wild birds for organic or conventional wheat (McKenzie and Whittingham, 2010). To carry out the study the team set up feeding stations in more than 30 gardens across the north of England. Organic and conventional wheat seeds (both cv. Alchemy) were placed in adjacent bird feeders and the rate at which the birds ate the different grain was monitored in three experiments over a 6-week period. Halfway through the experiment the position of the feeders was swapped. The experiment was repeated in a second winter with different wheat samples. The birds showed a strong preference for the conventional grain, eating significantly more of this than the organic grain. When the feeder positions were switched, the birds learned the new position of the conventional seed and continued to select it in preference to the organic. Analysis showed that the conventionally grown grain had a consistently higher protein content (by 10 percent) than the organic grain. The 1000-seed weight was also higher in conventional than in organic grain in two experiments, and was not measured in the third. Other differences between the samples (mycotoxin levels, energy content and pesticide residues) could not explain the preferences shown by the birds. The garden-bird work was confirmed by laboratory studies on canaries, also showing a significant preference for conventionally over organically grown grain. Neither *E. coli* nor *Salmonella* spp. were found in any of the feed samples. Enterobacteriaceae were detected in small quantities in one of the conventional samples used in one of the experiments. However, as this conventional type remained preferred to the uncontaminated organic grain, this is not likely to explain the difference in grain preference. The most likely explanation for the findings is that the birds found the conventional grain to be superior as a winter feed.

Oats

Few published data are available on the nutritional quality of organic oats. Dimberg *et al.* (2005) reported no significant differences in the concentration of phenolic compounds in conventional and organic oats. The compounds tested included avenanthramides (AVAs), hydroxycinnamic acids (HCAs), and a sucrose-linked truxinic acid (TASE), in three cultivars of oats (i.e. Freja, Sang and Matilda) grown in Sweden. Overall, there were significant differences between years, cultivars and fertilizer rate. The content of some of the phenolic compounds was correlated negatively with yield and specific weight.

Barley

Starling and Richards (1990) reported that samples from Scottish organic barley trials were of lower specific weight and 1000-grain weight than conventional trial samples. This was mainly due to mildew infection. The lower protein content was considered desirable for malting barley.

Conclusions

Based on the documented evidence, it does not appear that chemical and pesticide residues in conventional cereals are a major health concern for consumers. A more definite area of concern is the possibility of mycotoxin contamination, in both conventional and organic cereals.

Organic cereals generally have a lower content of protein, but in other nutritional aspects are similar to conventional cereals except for a slightly higher content of phenolic compounds. Animals appear to have the ability to discriminate between organic and conventional sources of cereals. In preference tests, rats selected biscuits made from organic wheat over biscuits made from conventional wheat. Conversely, wild birds preferred conventional wheat seed as a winter feed, a result ascribed to a lower content of protein in the organic wheat.

References

Anselme, M., Tangni, E.K., Pussemier, L., *et al.* (2006). Comparison of ochratoxin A and deoxynivalenol in organically and conventionally produced beers sold on the Belgian market. *Food Additives and Contaminants* **23**, 910–918.

Australian Department of Agriculture, Fisheries and Forestry (2010). Plant Products: Random Residue Monitoring Results. http://www.daff.gov.au/agriculture-food/nrs/programme, accessed September 5, 2010.

Biffi, R., Munari, M., Dioguardi, L., *et al.* (2004). Ochratoxin A in conventional and organic cereal derivatives: a survey of the Italian market, 2001–02. *Food Additives and Contaminants* **21**, 586–591.

Brookes, G. (2008). The impact of using GM insect resistant maize in Europe since 1998. *International Journal of Biotechnology* **10**, 148–166.

Cirillo, T., Ritieni, A., Visone, M. and Cocchieri, R.A. (2003). Evaluation of conventional and organic Italian foodstuffs for deoxynivalenol and fumonisins B-1 and B-2. *Journal of Agricultural and Food Chemistry* **51**, 8128–8131.

Dimberg, L.H., Gissen, C. and Nilsson, J. (2005). Phenolic compounds in oat grains (*Avena sativa* L.) grown in conventional and organic systems. *Ambio* **34**, 331–337.

Edwards, S.G. (2009). Fusarium mycotoxin content of UK organic and conventional oats. *Food Additives and Contaminants* A **26**, 1063–1069.

Eurola, M., Hietaniemi, V., Kontturi, M., *et al.* (2003). Cadmium contents of oats (*Avena sativa* L.) in official variety, organic cultivation, and nitrogen fertilization trials during 1997–1999. *Journal of Agricultural and Food Chemistry* **51**, 2608–2614.

Gans, W., Beschow, H. and Merbach, W. (2000). Growth regulators for cereal and oil crops on the basis of 2, 3-dichloroisobutyric acid and chlormequat chloride and residue analyses of both agents in the grain of oat. *Journal of Plant Nutrition and Soil Science* **163**, 405–410.

Haglund A., Johansson, L. and Dahlstedt, L. (1998). Sensory evaluation of wholemeal bread from ecologically and conventionally grown wheat. *Journal of Cereal Science* **27**, 199–207.

Hoogenboom, L.A.P., Bokhorst, J.G, Northolt, M.D., *et al.* (2008). Contaminants and microorganisms in Dutch organic food products: a comparison with conventional products. *Food Additives and Contaminants* **25**, 1197–1209.

Hornick, S.B. (1992). Factors affecting the nutritional quality of crops. *American Journal of Alternative Agriculture* **7**, 63–68.

Jestoi, M., Somma, M.C., Kouva., M., *et al.* (2004). Levels of mycotoxins and sample cytotoxicity of selected organic and conventional grain-based products purchased from Finnish and Italian markets. *Molecular Nutrition and Food Research* **48**, 299–307.

Juhler, R.K. and Vahl, M. (1999). Residues of chlormequat and mepiquat in grain – results from the Danish National Pesticide Survey. *Journal of AOAC International* **82**, 331–336.

Mäder, P., Hahn, D., Dubois, D., *et al.* (2007). Wheat quality in organic and conventional farming: results of a 21 year field experiment. *Journal of the Science of Food and Agriculture* **87**, 1826–1835.

McKenzie, A. and Whittingham, M. (2010). Birds select conventional over organic wheat when given free choice. *Journal of the Science of Food and Agriculture* **90**, 1861–1869.

Paepens, C., De Saeger, S., Sibanda, L., *et al.* (2005). Evaluation of fumonisin contamination in cornflakes on the Belgian market by "flow-through" assay screening and LC-MS/MS analyses. *Journal of Agricultural and Food Chemistry* **53**, 7337–7343.

Ryan, M., Derrick, J. and Dann, P. (2004). Grain mineral concentrations and yield of wheat grown under organic and conventional management. *Journal of the Science of Food and Agriculture* **84**, 207–216.

Starling, W. and Richards, M.C. (1990). Quality of organically grown wheat and barley. *Aspects of Applied Biology* **25**, 193–198.

Starling, W. and Richards, M.C. (1993). Quality of commercial samples of organically-grown wheat. *Aspects of Applied Biology* **36**, 205–209.

USDA (2007). US Department of Agriculture Pesticide Data Program, annual summary, calendar year 2006. USDA, Washington, DC. http://www.ams. usda. gov/pdp, accessed July 5, 2010.

USDA (2010). US Department of Agriculture Pesticide Data Program. http://www.ams.usda.gov/pdp, accessed April 26, 2010.

Woese, K., Lange, D., Boess, C. and Bogl, K.W. (1997). A comparison of organically and conventionally grown foods – results of a review of the relevant literature. *Journal of the Science of Food and Agriculture* **74**, 281–293.

Zenor, R.T. (2008). Incidence after Chewing Organic Rye Leads to Arrest. www. zenors.com/Pavillions/Incident.doc, accessed November 7, 2008.

Zörb, C., Niehaus, K., Barsch, A., *et al.* (2009). Levels of compounds and metabolites in wheat ears and grains in organic and conventional agriculture. *Journal of Agricultural and Food Chemistry* **57**, 9555–9562.

Zuchowski, J., Kapusta, I., Szajwaj, B., Jończyk, K. and Oleszek, W. (2009). Phenolic acid content of organic and conventionally grown winter wheat. *Cereal Research Communications* **37**, 189–197.

6 Meat

Among the concerns that consumers have about meat are possible chemical, antibiotic and hormone residues, "mad-cow disease", and that the meat may have come from cloned or gene-modified animals. As a result, some consumers choose organic meat over regular (conventionally produced) meat.

In the production of organic meat the animal must be given a prescribed amount of living space (no cages) and access to the outdoors and must be fed organic feed. No antibiotics, growth hormones or other artificial drugs can be used. If an animal is diseased and requires medication, it can be treated but can no longer be considered organic. Production of organic meat is more expensive than on conventional farms because it is more labor-intensive, the supply of organic feedstuffs is limited, the animals tend to grow more slowly, production is affected more by climate than on conventional farms which provide housing and shelter, and organic production is conducted mainly on smaller farms that do not offer the economies of scale available on larger farms.

Consumers generally assume that organic meat is superior to meat produced conventionally. It is interesting to review the research findings relating to this issue to determine whether the concerns about conventional meat are justified.

Chemical and pesticide residues

In the mid-1990s the Food Safety and Inspection Service of the US Department of Agriculture and the Environmental Protection Agency conducted a survey to gather information on dioxin-like chemicals in beef, pork and poultry products. The purpose was to measure the extent of potential contaminants in the food supply.

Low levels of dioxin-like chemicals are common in the environment. They are released into the environment through natural processes such as forest fires and volcanic eruptions, and through industrial processes including combustion or incineration of industrial waste or chemical manufacturing. These compounds can remain in the environment for decades.

Organic Production and Food Quality, First Edition. Robert Blair.
© 2012 John Wiley & Sons, Ltd. Published 2012 by John Wiley & Sons, Ltd.

Dioxin-like chemicals accumulate in the fatty tissues of food animals, and the primary means of human exposure is believed to be through the consumption of animal fats in food. They also accumulate in fatty tissue in the human body. Studies have shown that prolonged exposure to elevated levels of dioxin may have long-term, adverse health effects. Animal feed contaminated with dioxins was reported in Ireland in late 2008, causing the authorities to take swift and effective remedial action.

In the survey conducted by the Food Safety and Inspection Service of the US Department of Agriculture and the Environmental Protection Agency, samples of fatty tissue were collected from carcasses of 51 steers/heifers, 56 market pigs, 41 young chickens, and 15 young turkeys. Analyses for dioxin-like chemicals showed that most levels were below 2.0 parts per trillion (ppt). Some were barely at the level of detection. Two of 41 young chicken samples had highly elevated levels (25–31 ppt) and it was subsequently found that a type of clay added to the feed as an anti-caking agent was the cause of the elevated levels. As a result the Food and Drug Administration banned the use of that type of anti-caking agent in animal feeds.

The survey was repeated by the US authorities in 2002–2003, to update the information. Twenty different dioxin-like chemicals were analyzed in samples collected from carcasses of 510 market pigs, steers/heifers, young chickens, or young turkeys processed in federally inspected slaughter establishments. Results are shown in Table 6.1.

These findings indicate that the meat supply in the USA is not heavily contaminated by dioxins and similar environmental contaminants, and that the efforts being made to reduce the levels appear to be succeeding. The levels found in the survey can be regarded as trace amounts.

The latest available results for the 2009 year and published in 2011 from the Food Safety and Inspection Service of the US Department of Agriculture confirmed these findings. In total 17 241 samples were tested for all residues required under the testing program. The number of violations totaled 21 (0.12 percent, Table 6.2), due mainly to residues of carbadox (anti-microbial agent), florfenicol (anti-microbial agent), and sulfas (anti-microbial agents).

No residues were detected in approximately 97 percent of the domestic samples. The program reported 21 residue violations (0.12 percent) in

Table 6.1 Levels of dioxin-like chemicals (ppt, parts per trillion) found in meat in 1994–1996 and 2002–2003 (USDA, 2009; www.fsis.usda.gov/PDF/Dioxin_ Report_0605.pdf)

Slaughter class	1994–1996 ppt	2002–2003 ppt
Market pigs	1.47	0.28
Steers/heifers	1.38	0.93
Young chickens	0.94	0.33
Young turkeys	1.53	0.63

Table 6.2   Incidence of residues in meat (USDA, 2011; http://www.ams.usda.gov/pdp)

Compound class	Number of samples	Number of non-violative positives	Number of violations	Percent violations
Antibiotics	5 154	334	5	0.10
Arsenic	1 473	84	0	0.00
Avermectins	1 645	27	2	0.12
Beta-agonists	372	2	0	0.00
Carbadox	372	3	2	0.54
Chloramphenicol	1 369	0	0	0.00
Florfenicol	426	0	4	0.94
Flunixin	579	0	0	0.00
Furazolidone	644	0	1	0.16
Nitroimidazoles	633	0	0	0.00
Pesticides	1 268	23	1	0.08
Sulfas	2 496	0	6	0.24
Thyreostats	216	0	0	0.00
Trenbolone	448	0	0	0.00
Zeranol	146	0	0	0.00
Total	17 241	473	21	0.12

the following: one beef cow, two bob veal, two bulls, one dairy cow, one formula-fed veal, one goat, one heavy calf, one market hog, five non-formula fed veal, four roaster pigs, and two steers. Of the 21 residue violations, six were attributed to sulfas, five were attributed to antibiotics, two each were attributed to avermectins and carbadox, four were attributed to florfenicol, and one each were attributed to nitorfurans, and pesticide, respectively.

Australia has a very good food monitoring program and publishes a detailed report annually of the results. The most recent report (DAFF, 2010) is for the 2008–2009 testing period (Table 6.3). During that period the monitoring projects covered 12 meat commodities that included 15 426 samples. Cattle, sheep and pigs provided the largest numbers of samples (5732; 5475 and 3302 samples respectively). Samples complying with the standards were almost 100 percent (and it was a relief to see that kangaroo samples were 100 percent in compliance!). Non-compliance was found in 21 samples (relating to 10 pesticide and veterinary medicinal residues). Eleven environmental contaminants were found at concentrations above applicable Australian Standards. One cattle and four sheep liver samples had cadmium levels that exceeded the Australian Standard maximum level (ML) of 1.25 mg/kg. Five sheep and one wild boar liver samples (six in total) contained lead residues higher than the ML of 0.5 mg/kg. Follow-up action was taken in all of these cases.

Table 6.3   Results of Australian Food Monitoring Program for chemical and pesticide residues in meat (2008–2009)

Meat type	Total samples	Residues exceeding standards	Samples compliant with standards (%)
Camel	4	0	100
Cattle	5 732	2	99.95
Deer	67	0	100
Goat	250	2	99.2
Horse	145	0	100
Kangaroo	55	0	100
Pig	3 302	3	99.91
Poultry	330	0	100
Ratite (emu)	7	0	100
Ratite (ostrich)	4	0	100
Sheep	5 475	2	99.8
Wild boar	55	2	96.36
Total	15 426	10	99.86
Farmed fish	15	0	100
Wild fish	213	0	100

NA = Not applicable.
(*Source*: DAFF (2010). Australian National Residue Survey 2008–2009. www.daff.gov.au/agriculture-food/nrs/animal2. Reprinted with permission of the Department of Agriculture, Fisheries and Forestry.)

Australian Standards (maximum residue limits (MRLs) and maximum levels (MLs)) apply only to residues found in edible tissues. Some residue testing is undertaken in non-edible tissues for monitoring purposes because residues of particular chemicals in these tissues can be an indication of illegal or inappropriate use of those chemicals.

The results in Table 6.3 show that Australian meat is of high quality, with very few samples containing residues that exceed the national standard.

Results from Ireland indicate that Irish meat also is of high quality, with very few samples containing residues that are in non-compliance with or exceed the national standard (Table 6.4 and Table 6.5). Over the period 1998–2005, the levels of chemical residues in beef, pork and chicken in that country have dropped, although milk has shown less of a sustained drop. The results for mutton/lamb show a reverse trend: a slight increase over the period. Mycotoxins (fungal toxins) were found in meat from cattle, sheep and poultry, also in honey and milk, indicating some contamination of the feed or perhaps bedding.

The most recent results from this agency (Irish National Food Residue Database, 2009) indicate that residues of the common pesticides are not a problem in foods of animal origin. The number of samples analyzed for the

Table 6.4 Results of Irish Food Monitoring Program for chemical residues and mycotoxins in meat (2005). See Ireland National Food Residue Database (2009)

Commodity	Chemical residues		Mycotoxins	
	Total samples	Samples non-compliant	Total samples	Samples non-compliant
Cattle	141	0	31	0
Sheep	34	0	9	0
Pig	38	0	13	0
Poultry	28	0	18	0
Goat	29	0	Not tested	Not tested
Horse	5	0	Not tested	Not tested
Milk	140	0	69	0
Honey	Not tested	Not tested	11	0
Eggs	Not tested	Not tested	Not tested	Not tested

(*Source*: adapted from Danaher, M., Sherry, A.M. and O'Mahony, J. National Food Residue database report 2009. Teagasc Food Research Centre, Ashdown, Dublin, Ireland (2010). Copyright Teagasc 2010.)

years 2002 to 2007 ranged between 57 and 130 per year for beef fat, 48 and 83 per year for mutton fat, 52 and 86 per year for pig fat, 50 and 67 per year for dairy produce, 0 and 24 per year for venison, and 23 and 45 per year for poultry meat. The samples of meat, venison, poultry and dairy produce were all of domestic origin. The samples of food of animal origin were

Table 6.5 Results of Irish Food Monitoring Program for hormone-type and pesticide-type residues in meat (2005). See Ireland National Food Residue Database (2009)

Commodity	Hormone-type		Pesticide-type	
	Total samples	Samples non-compliant	Total samples	Samples non-compliant
Cattle	2714	6	6054	35
Sheep	476	0	1531	6
Pig	751	0	25518	26
Poultry	546	0	2552	16
Goat	10	0	72	0
Horse	67	0	86	0
Milk	249	0	761	2
Honey	20	0	71	0
Eggs	95	0	183	0

(*Source*: adapted from Danaher, M., Sherry, A.M. and O'Mahony, J. National Food Residue database report 2009. Teagasc Food Research Centre, Ashdown, Dublin, Ireland (2010). Copyright Teagasc 2010.)

tested for approximately 55 pesticides and metabolites, including seven PCB congeners in each year. No residues of pesticides were detected in 67 to 100 percent of the total number of foods of animal origin samples tested in each year and, of the samples for which detectable pesticide residues were determined, in only two cases were the relevant MRLs exceeded in 2006. The pesticides detected in foods of animal origin were primarily the organochlorine pesticides, DDT, hexachloroben-zene, dicofol, lindane and dieldrin, PCB congeners and the organophos-phorus pesticide, diazinon. The organochlorine pesticides and PCB congeners are associated with background levels in soil or feedingstuffs and the diazinon residues in fat may be associated with use of an ecto-parasiticidal treatment.

In a research study Heaton *et al.* (1996) tested samples of ham, pork sausage, pork carcass (adipose and muscle tissues), pork liver and pork kidney and found no violative residues of 25 different chlorinated-hydrocarbon/organophosphate pesticides and/or of seven different European Community (EC)-specified compounds/compound classes.

Organic meat

An important question raised by consumers is whether organic meat contains fewer residues than conventional meat. Although there is a dearth of information in the scientific literature on levels of chemical residues in organic meats, a few comparisons of organic and conventional beef have been conducted.

Smith *et al.* (1994) collected samples of beef from steers, heifers and cows at eight packing plants in four states in the USA. The aim of the study was to determine whether US beef met the standards set by the European Community. The animals included "conventional", "natural" and "organic" cattle. The USDA approved use of the designation "natural" in 1982 to indicate beef that has been minimally processed and contains no artificial flavoring, coloring, chemical preservatives or other synthetic ingredients. The study involved the testing of 20 samples each of muscle, adipose tissue, kidney and liver for residues of diethylstilbestrol, zeranol, trenbolone acetate and melengestrol acetate (anabolic steroids); lead and cadmium (environmental contaminants); carazolol (a beta-blocker); clenbuterol (a beta-agonist); azaperone and propiopromazine (tranquillizers) and six sulfa drugs: sulfamethazine, sulfadimethoxine, sulfabromomethazine, sulfa-ethoxypyridazine, sulfachloropyridazine, and sulfamethoxypyridazine. In addition, adipose (fatty) tissue samples from 20 carcasses in each category were tested for residues of 25 chlorinated-hydrocarbon and organophos-phate pesticides. In all samples no residue amount that would be considered violative was detected and it was concluded that US beef met the standards for importation into Europe.

Table 6.6 Incidence of chemical residues (µg/g) in samples of muscle tissue of conventional, natural and organic beef animals (Smith *et al.*, 1994; © John Wiley & Sons)

	Type of beef		
	Conventional	Natural	Organic
Compound wet-wt basis			
Carazolol	<0.001	<0.001	<0.001
Clenbuterol	<0.001	<0.001	<0.001
Azaperone	<0.004	<0.004	<0.004
Propiopromazine	<0.001	<0.001	<0.001
Compound (dry-wt basis)			
Diethlystilbestrol	<0.001	<0.001	<0.001
Zeranol	<0.001	<0.001	<0.001
Trenbolone	<0.001	<0.001	<0.001
Melangestrol	<0.001	<0.001	<0.001
Sulfamethazine	<0.025	<0.025	<0.025
Sulfadimethoxine	<0.025	<0.025	<0.025
Sulfabromomethazine	<0.025	<0.025	<0.025
Sulfaethoxypyridazine	<0.025	<0.025	<0.025
Sulfachloropyridazine	<0.025	<0.025	<0.025
Sulfamethoxypyridazine	<0.025	<0.025	<0.025
Lead	0.358	0.838	1.627
Cadmium	0.048	0.042	0.014

Results for chemical residues in muscle tissue are shown in Table 6.6 and for pesticide residues in Table 6.7. The lead content was higher in organic beef and the cadmium level was higher in conventional beef, but neither difference achieved statistical significance.

None of the differences observed achieved statistical significance.

The same group (Smith *et al.*, 1997) at the Center for Red Meat Safety and the Warren Analytical Laboratory in Colorado repeated the study and analyzed a total of 186 samples of muscle, adipose, liver and kidney tissues from beef animals labeled conventional, natural or organic.

There were no violative residues (Table 6.8) of anabolic steroids (estrus suppressants, growth promotants), xenobiotics (growth promotants), sulfa drugs (growth promotants, health aids), antibiotics (health aids) or tetra-cyclines (health aids). When violative residues occurred, the residues were of pesticides, and the highest incidence was in livers from beef cattle produced under "natural" (6 of 1575 tests; 0.38 percent) and "organic" (6 of 1575 tests, 0.38 percent), followed by "conventional" (3 of 1500 tests) management conditions. The only violative residues of any chemical found in these studies were in livers and not in meat *per se*. Results of this study

Table 6.7 Incidence of pesticide residues (μg/g) in samples of adipose tissue of conventional, natural and organic beef animals (Smith *et al.*, 1994; © John Wiley & Sons)

Compound wet-wt basis	Tolerance allowed	Conventional	Natural	Organic
Hexachlorobenzene	0.500	0.002	0.002	<0.0030
Lindane	7.000	<0.002	<0.002	<0.002
Heptachlor	0.200	<0.002	<0.002	<0.002
Aldrin	0.300	<0.002	<0.002	<0.002
4,4'-DDT	5.000	<0.002	<0.002	<0.002
4,4'-DDD	5.000	<0.002	<0.002	<0.002
4,4'-DDE	5.000	0.0049	0.0066	0.0077
Endrin	0.300	<0.002	<0.002	<0.002
Mirex	0.100	<0.002	<0.002	<0.002
Ethyl parathion	0.100	<0.020	<0.002	<0.020
Methyl parathion	0.100	<0.020	<0.002	<0.020
Pirimiphos-methyl	0.200	<0.020	<0.002	<0.020
Alpha-BHC	0.300	<0.0021	<0.002	<0.002
Beta-BHC	0.300	<0.002	<0.002	<0.002
Delta-BHC	0.300	<0.002	<0.002	<0.002
Heptachlor epoxide	0.200	<0.002	<0.002	<0.002
Methoxychlor	3.000	<0.002	<0.002	<0.002
Ethion	2.500	<0.020	<0.020	<0.020
Chlorpyrifos	2.000	<0.020	<0.020	<0.020
Malathion	4.000	<0.020	<0.020	<0.020
Ronnel	10.000	<0.020	<0.020	<0.020
Trithion	0.100	<0.020	<0.020	<0.020
Dieldrin	0.300	0.0033	0.0020	0.0027
Diazinon	0.700	<0.0205	<0.020	<0.020
Disyston	0.100	<0.020	<0.020	<0.020

confirm that it is highly unlikely there is any difference in occurrence of chemical residues of drugs, vaccines, pesticides, antibiotics and/or growth promotants in regular, natural or organic beef.

Some of the differences found in residue level between beef type were statistically significant. Livers from conventional beef animals had a higher content of α-BHC, 4,4'-DDE and methyl parathion than livers from organic beef animals, while livers from organic beef animals had a higher content of diazinon. Natural beef animals had a higher content of 4-4'-DDE in the kidney, endrin in the liver, and diazinon in the liver than conventional beef animals. The differences were regarded as minimal.

This was a detailed study, as was the previous study by these researchers. The chlorinated hydrocarbon pesticides studied included aldrin, BHC

Table 6.8 Incidence of chlorinated-hydrocarbon and organophosphate pesticides in samples of conventional natural and organic beef (Smith *et al.*, 1997; © John Wiley & Sons)

Beef type	Tissue	Samples analyzed	Chlorinated-hydrocarbon pesticides		Organophosphate pesticides	
			Samples with violative residues	Samples with detectable but non-violative residues	Samples with violative residues	Samples with detectable but non-violative residues
Conventional	Muscle	20	0	2	0	6
	Adipose	20	0	6	0	4
	Liver	10	3	10	0	1
	Kidney	10	0	0	0	1
Natural	Muscle	20	0	1	0	2
	Adipose	20	0	7	0	2
	Liver	13	3	5	3	0
	Kidney	10	0	3	0	2
Organic	Muscle	24	0	0	0	0
	Adipose	20	0	3	0	0
	Liver	13	3	4	3	0
	Kidney	6	0	0	0	0

isomers, DDT and metabolites, dieldrin, mirex, endrin, HCB, heptachlor, heptachlor epoxide, lindane and methoxychlor. The organophosphate pesticides studied included diazinon, methyl parathion, ronnel and chlorpyrifos. No violative residues of anabolic steroids (testosterone; estradiol; progesterone), xenobiotics (zeranol; melengestrol acetate; trenbolone acetate), penicillin; macrolide antibiotics (tylosin, erythromycin), sulfa drugs (sulfathiazole; sulfamethazine; sulfadimethoxine; sulfaquinoxaline), tetracycline antibiotics (tetracycline; oxytetracycline; chlortetracycline) or pesticides of the chlorinated hydrocarbon and organophosphate groups (lindane; heptachlor; aldrin; 4,4'-DDT; 4,4'-DDD; 4,4'-DDE; ethyl parathion; methyl parathion; pirimiphos-methyl; alpha-BHC; beta-BHC; delta-BHC; heptachlor epoxide; rnethoxychlor; ethion; chlorpyrifos; malathion; ronnel; trithion; dieldrin; disyston).

There were six violative residues of pesticides in livers from organic beef (three of hexachlorobenzene; three of diazinon), six violative residues of pesticides in livers from natural beef (two of hexachlorobenzene; one of endrin; three of diazinon) and three violative residues of pesticides in livers from conventional beef (two of hexachlorobenzene; one of mirex).

If the samples are typical of beef in the current national supply, the data from these studies indicate that the incidence of violative chemical residues in US beef is very low.

In related work, Usborne (1994) compared natural and conventional beef purchased in retail supermarkets in Canada and reported no violative residues of sulfa drugs, antibiotics, heavy metals, polychlorinated biphenyls, growth promotants, parasiticides, pentachlorophenol (a wood fungicide) or pesticides in either kind of beef.

Research was carried out with pigs at the Swedish University of Agricultural Sciences to determine the effects of organic rearing on kidney levels of cadmium, a toxic trace mineral (Lindén et al., 2001). The conventional pigs were raised indoors and the organic pigs were raised outdoors with access to soil. The conventional diet contained 51.8 μg/kg (ppb) cadmium and the organic diet 39.9 μg/kg (ppb). A surprising finding of that study was that it was the organic pigs that had the higher level of cadmium in the kidney, 84.0 μg/kg (conventional) vs. 96.2 μg/kg (organic) although the level of this mineral was lower in the organic diet. In trying to explain the results the researchers suggested that the uptake of cadmium in the gut may have been higher with the organic diet. Another possible explanation was the intake of cadmium from soil. Other research has shown that the intake of soil by pigs may be about 8 to 12 percent of dietary intake. However, based on calculations of cadmium intake from soil and feed, the researchers calculated that the cadmium intake of the organic pigs would have been similar to that of the conventional pigs. Therefore other factors appeared to be involved in causing the higher kidney cadmium levels in the organic pigs. There was no significant difference in liver cadmium levels between organic and conventional pigs. The level in pork muscle was not measured.

Blanco-Penedo et al. (2010) measured the content of trace elements in beef (diaphragm muscle) of cattle raised under organic, intensive or conventional production systems. Cadmium concentrations were low (<10 μg/kg wet weight) and muscle arsenic, mercury and lead levels were below the limits of detection (<12, 2 and 3 μg/kg, respectively) in most samples (77–97 percent). There were no significant differences between farms. Essential trace element concentrations in muscle were generally within adequate physiological ranges and, although they varied significantly between farms, this did not appear to be related to management practices.

The overall conclusion that can be drawn from the available data is that the level of chemical and pesticide residues in regular (conventionally produced) meat should not be a major concern for the consumer. Those who have a concern because of this issue should trim off the fat from the meat, since some pesticides accumulate in the fat. In addition they should avoid organ meats such as liver and kidney since these organs are associated with the elimination of residues from the animal body. Consumers not convinced by the available evidence should purchase organic meat, though it is clear that organic meat is not completely devoid of residues.

Some monitoring authorities test organ tissues rather than meat samples since the results are usually a better indication of pesticide exposure than muscle tissue.

Hormones

The public has an ambivalent attitude towards hormones. On the one hand they feel they are "gender-benders" and to be feared, but on the other hand they are quite happy to take them for hormone-replacement therapy, birth control, sleep, to build bone and muscle, for anti-aging, skin conditions, and so on. The presence of hormones in foods is a controversial topic.

Hormones are naturally occurring substances in animals and plants. This means that humans not only produce hormones but also ingest them when they consume foods such as meat, milk, cabbage and soybeans. The amount of hormones provided by the diet, however, is much lower than the amount produced in the human body. For instance, one cup of milk contains on average 35.9 ng of estrogen (0.000 000 001 g), compared with the 136 000 ng estrogen (3800 times greater) produced naturally each day in men, and the 192 000–1 192 000 ng (5400 times greater) in non-pregnant women.

There are two types of hormones, steroid and protein. Steroid hormones are active in the body when taken in by mouth. For example, birth control pills are steroid hormones and can be taken orally. Bovine somatotropin (bST, which is used to stimulate milk production in dairy cows) is a protein hormone and has no activity when taken in by mouth because the body digests it like other proteins. It has to be injected to be active. Insulin is another protein hormone that is inactive when taken in by mouth, which explains why insulin-dependent diabetics have to inject it.

The history of hormone use in agriculture has been tainted by the diethyl stilbestrol (DES) saga. This synthetic hormone was first synthesized in 1938 and became used for the prevention of miscarriages and premature deliveries in several million pregnant women in North America and Europe. Unfortunately DES was later shown to be a carcinogen, causing problems in the daughters of women who had taken it. It was first approved for use by the FDA in 1954 for addition to cattle feed because it stimulated growth and improved productivity. Later in 1956 it was approved as a subcutaneous implant for steers. As a result of the medical problems it was taken out of the food chain and banned from feed or implants in cattle and sheep in the United States in 1979. It was taken off the Canadian market in 1971 and is now banned for agricultural use in most countries. However, it should be noted that the link between its prescribed use in mothers and the incidence of a rare form of cancer in daughters did not extend to an established link between DES use in cattle and sheep and the cancer problem in humans. Nevertheless that experience has tainted the image of hormone use in agriculture and resulted in the understandable apprehension felt today about hormones in general.

DES has not completely disappeared from the market. It is still being prescribed by physicians for prostate cancer. Also, it can be obtained over the internet from some European countries.

Various comments in the media and on the internet indicate that there is a certain element of public misinformation about hormones and food. For instance a common myth is that chickens and turkeys are fed or injected with hormones to make them grow fast. The reality of the situation is that no poultry company in the world does this, as reported by Dr Tim Cummings of the College of Veterinary Medicine at Mississippi State University (Cummings, 2011).

So why do modern chickens grow so fast, if not "stuffed with hormones" as claimed by some commentators? Scientists at the University of North Carolina (Havenstein *et al.*, 2003) studied this question and, after analysis of the appropriate data, concluded that genetic selection by commercial breeding companies had brought about 85 to 90 percent of the change seen in broiler growth rate over the past 45 years. A major influence was the use of hybrids instead of pure breeds. Hybrids (crossings of two or more breeds) show "hybrid vigor", with production characteristics superior to those of the parent purebreds. Better nutrition made up the remaining 10 to 15 percent of the improvement seen over the previous 45 years.

Two main processes occur in the animal (and human) body. Anabolism involves the synthesis of protein tissue and muscle growth. The opposite is catabolism, i.e. breakdown and loss of muscle tissue. Hormones, natural and synthetic, can be used to regulate these processes and give an increased content of muscle tissue – meat – in the carcass and faster growth. Consumer demand is now for meat with less fat and more lean. Fast-growing animals deposit less fat in the carcass, therefore the meat is leaner.

There are six hormones approved for use in beef production in more than 30 countries, following reviews by regulatory authorities on their safety and efficacy. Three of these are natural, three synthetic. The three natural hormones (testosterone, estradiol, and progesterone) have been deemed completely safe for use in beef production since they are a natural part of all mammalian physiology. Their use is non-controversial. The three synthetic growth-enhancing hormones are melengestrol acetate (MGA), trenbolone acetate (TBA), and zeranol. These are more stable forms of the natural hormones.

Zeranol has been approved as a subcutaneous ear implant in cattle and sheep. In steers (young male castrates) it increases protein retention, growth rate and efficiency of feed conversion. Lower responses are seen in heifers (young females). Each implant contains a specific, authorized dose of the hormone. The implant ensures that the hormone is released into the bloodstream very gradually over time so that the concentration of the hormone in the animal remains relatively constant and low. All of these products have undergone extensive eco-safety assessments, including worst-case estimates of their levels in cattle manure, runoff from cattle feedlots, and runoff from land on which the manure has been applied. The ears are discarded when the animal is slaughtered, so that any residue remaining does not enter the food chain.

A research team at the University of Hamburg in Germany studied the natural occurrence of sex steroid hormones in foods, i.e. progesterone, testosterone, 17β-estradiol and estrone (Hartmann *et al.*, 1998). Related compounds such as DHEA (dehydroepiandrosterone, thought to have anti-aging properties) were also measured. It was found that the steroid patterns of beef, pork, other meat products, fish and poultry were similar. Milk and milk products reflected the hormone profile of female cattle, with higher amounts of progesterone in milk with an increased content of milk fat. Milk products supplied about 60–80 percent of female sex steroids present in the diet. Eggs were a considerable source of all of the steroids investigated, about the same as meat and fish (10–20 percent). No estrogens could be detected in vegetable food. However, plants supplied similar amounts of testosterone as meat and milk products (20–40 percent). They also contained considerable amounts of hormone precursors (contributing about 80 percent to the DHEA supply). These researchers concluded that, compared with the human daily production of steroid hormones (progesterone equal to about $10\,\mu g/day$, testosterone $0.05\,\mu g/day$, estrogens $0.1\,\mu g/day$, DHEA $0.5\,\mu g/day$), the amount in the diet is insignificant.

No hormones are used in chicken or swine production since they are less effective in these species, therefore the question of elevated hormone residues in these meats does not arise.

Hormone residue levels in meats. Do government agencies monitor for these?

Estradiol, progesterone and testosterone are sex hormones that are synthesized naturally in the animal body. No regulatory monitoring of these hormones is possible, since it is not possible to identify separately the hormones administered and those made by the animal naturally. However, it is possible to detect residues of zeranol and trenbolone acetate in the meat. The US Food and Drug Administration (FDA) has set the tolerance levels for these hormones, i.e. the maximum amount of a residue that is permitted in food. The Food Safety Inspection Service (FSIS) of the US Department of Agriculture (USDA) has the responsibility for monitoring meat.

All of the official bodies that have assessed the scientific evidence have concluded that the use of approved hormones in beef and milk production present no harm to the consumer. Foods from treated animals have been deemed safe. These bodies include the FDA and USDA in the United States, Health Canada, the World Health Organization (WHO), other European scientific bodies and the Food and Agriculture Organization of the United Nations (FAO).

The Joint Expert Committee on Food Additives of the World Health Organization and FAO (WHO/FAO Expert Committee) calculated that, even assuming the highest residue levels found in beef, a person consuming

1 lb (about 500 g) of beef from an implanted steer would take in only 50 ng of additional estradiol compared with non-implanted beef. That is less than one-thirtieth of the acceptable daily intake (ADI) of estradiol for a 75 lb (34 kg) child established by the WHO/FAO Expert Committee.

The United States, Canada and Australia are among countries that permit the use of hormone implants in cattle, deeming them safe and as presenting no danger to the consumer of meat from treated animals. However, the European Union (EU) does not agree and currently bans the importation of meat from animals that have received hormone implants. This disagreement has trading implications. By banning implants in beef cattle, the EU can prevent the import of beef from countries that allow this technology. Despite claims made by the EU that there is a strong scientific basis for the ban, the World Trade Organization has ruled that these health concerns remain unproven and that the ban is not justified under its rules.

Renu Gandhi and Suzanne M. Snedeker of the Program on Breast Cancer and Environmental Risk Factors (BCERF) at Cornell University, Ithaca, New York, have summarized the relevant information on hormones in their factsheet "Consumer Concerns About Hormones in Food" and have concluded, "This fact sheet addresses some of the consumer concerns that have been brought to BCERF regarding health effects of hormones used by the meat and dairy industries. Evidence available so far, though not conclusive, does not link hormone residues in meat or milk with any human health effect." One reason for the caution expressed by these researchers is that long-term exposure to hormone-replacement therapy in post-menopausal women is associated with an increased risk of cancer.

Developing countries

The situation regarding hormone and chemical residues in foods is probably of more importance in developing countries that lack the regulatory controls on food production which developed countries rely on to ensure safe food. Research findings from these countries provide a guide for the selection and preparation of safe food from these countries, when we suspect that the food in question may be contaminated with hormones or residues.

For instance, an Egyptian study (Sadek et al., 1998) found that beef samples from cooperative supermarkets contained higher levels of trenbolone than meat from private butchers. The chemical was found in muscle, liver, kidney, fat and blood plasma. Beef and liver samples collected from government cooperative supermarkets and private butchers were free of zeranol. Levels of diethylstilbestrol (DES) and estradiol were also very low. The data also showed the level of estradiol was much higher in chicken from private growers than in chicken from government cooperative

Table 6.9 **Effect of cooking method on the level of estradiol (μg/kg) in Egyptian chicken meat and liver (Sadek *et al.*, 1998)**

	Muscle	Liver
Uncooked	0.280	0.757
Boiled with skin	0.160	0.625
Boiled without skin	0.020	0.625
Roasted with skin	0.210	0.762
Roasted without skin	0.100	0.762
Grilled with skin	0.190	0.721
Grilled without skin	0.060	0.721

supermarkets. It appeared that some poultry growers in Egypt were using hormones or hormone-like agents (such as oral contraceptive steroids) to improve the rate of growth. The researchers therefore extended the study to investigate whether the cooking method might affect the level of hormonal residue in chicken. They found (Table 6.9) that the best method for removing residue was boiling without the skin. Removal of the skin, which contains fat, contributed greatly to a decrease in estrogen intake by the consumer. The Egyptian study showed that consumers worried about possible intake of residues in chicken should grill or boil the skinless bird and avoid the fat and liver. These findings may help to explain why McDonald's restaurants in Egypt are so busy!

Antibiotics

Antibiotics may be used to treat sick animals that are being raised for meat. They may also be included in the diet to cope with a low level of exposure to disease organisms. As a result the animals may grow faster because of a better ability to fight off disease. This practice is permitted under the feeds regulations in various countries, but is being phased out.

Both the veterinary and farm use of antibiotics is under scrutiny because of the possible relationship to the emergence of antibiotic-resistant strains of bacteria affecting humans. As a result the use of antibiotics on farms is being increasingly curtailed, although it is more likely to be the veterinary use of antibiotics that is related to the antibiotic-resistance problem since the amounts administered to treat disease are much greater than those included in the diet. Another development is the increasing use of antibiotics that act only in the gut and that are not absorbed into the body and are therefore less likely to result in antibiotic resistance.

The public often wonders why antibiotics are being used on farms at all. The need to treat sick animals is understandable, but why include anti-biotics in the diet? The answer is that when penicillin was first produced commercially a use had to be found for the spent growth medium used to produce the penicillin. Somebody had the bright idea of feeding it to pigs, and lo and behold the pigs grew faster. No doubt this was due to the residual penicillin in the spent growth medium. So the practice began.

It is usually assumed that animals fed antibiotics grow faster because the antibiotic kills off disease organisms in the gut that depress growth, but this explanation is not quite complete. In the 1950s Dr Marie Coates, a researcher at the University of Reading in England, fed a diet containing antibiotic to chicks that had been raised germ-free. These chicks also grew faster. How is that result explained, since these chicks had no bacteria in the gut for the antibiotic to attack? A possible, alternative, explanation is that antibiotics act to thin the gut wall, allowing a more efficient absorption of nutrients.

The question of meat being contaminated with antibiotics is of secondary importance to the broader issue of human health being possibly compromised by antibiotic usage on farms, leading to the emergence of resistant strains of pathogenic bacteria.

Trace levels of antibiotics have been found in meat, as indicated above, but generally in tissue such as liver or kidney. One reason for the low levels found in meat is that most of the antibiotics now used in feed are not absorbed through the gut wall and do not get into the meat. The residues are excreted in the urine and feces.

Organic farming prohibits the routine use of antibiotics, and an antibiotic-treated animal loses its organic status. Thus the possibility of the emergence of antibiotic-resistant strains of bacteria is reduced or avoided on organic farms. But this raises another issue: since antibiotics are not routinely fed to organically raised animals, is it possible that these animals may harbor higher levels of pathogenic bacteria, especially those that cause food-poisoning?

Obviously, then, there are pros and cons to using antibiotics on farms. Organic farms that do not use antibiotics ought to pose less of a risk to human health from antibiotic resistance. On the other hand, animals have to be protected from developing a disease, and have to be treated when sick. Organic stock that are allowed to range outdoors also have to be protected from infection spread by wild birds and animals. Organic procedures need to be effective in this regard so as to minimize any risks to the health of the consuming public. Eventually vaccination of farm stock may prevent or reduce the risks from food poisoning from meat, provided vaccination is acceptable to the organic industry.

Since antibiotics are not routinely fed to organically raised animals, it has been suggested that these animals may harbor higher levels of pathogenic bacteria. On the other hand, bacteria present in these animals may be more susceptible to antibiotics because they have not developed resistance. Other factors, such as density of animals, confinement vs. free-range management, and hygiene practices, may also affect persistence and spread of bacteria in a flock or herd.

Some studies with dairy cattle have reported more antibiotic-resistant bacteria in animals that had been raised conventionally while others reported a similar incidence of resistance in bacteria from animals in both production systems (Doyle, 2006).

Bacterial contamination of meat

A concern that some consumers express about organic meat from animals raised outdoors is the possibility of bacterial contamination. *Campylobacter jejuni* is the most common enteric bacterial pathogen reported in developed countries and is considered to be of food-borne origin. Sporadic cases of *Campylobacter* infections during the summer months are mainly attributed to improper handling or consumption of undercooked poultry or the consumption of raw, unpasteurized milk or contaminated water. According to a study conducted by the USDA Food Safety and Inspection Service in 1994 to 1995, the prevalence of *Campylobacter* on immersion-chilled poultry carcasses was 88.2 percent (USDA, 1996).

Research data suggest that *Campylobacter* is primarily transferred on to poultry carcasses via fluid and excreta from the gastrointestinal tract of the bird, due to the high numbers of the organism found in these fluids. The organism then attaches to the skin and perseveres to final products. Davis and Conner (2000) reported that the incidence on raw, retail poultry products decreased from 76 percent on whole broilers to 48 percent on skin-on split breast and to only 2 percent on boneless, skinless breast meat.

The leading cause of human food-borne infections associated with consumption of poultry products world-wide is *Salmonella* (Van Immerseel *et al.*, 2002). Poultry can become infected from sources such as litter, droppings, soil, insects, rodent infestations, etc., and the most serious serotypes are those that can pass from the intestine of birds into the tissues to contaminate the meat and eggs. Prevention of infection by appropriate management protocols, including proper hygiene, is the most important control measure. Feed-related control measures that are known to be effective in helping to control *Salmonella* contamination levels include steam-pelleting of the feed, also the inclusion of approved additives such as antibiotics, probiotics and short-chain fatty acids in the feed mixture.

Information relating to organic meat was provided by Reinstein *et al.* (2009) who studied the prevalence of *E. coli* O157:H7 in beef cattle that had been raised organically and conventionally. This bacterium causes severe food poisoning in humans. In organically raised cattle, the average prevalence of *E. coli* O157:H7 in fecal samples was 9.3 percent (range 0 to 24.4 percent). In conventionally raised animals it was 6.5 percent. No major difference in antibiotic susceptibility patterns among the isolates was observed.

Another related finding was that when cattle naturally infected with *E. coli* O157:H7 were abruptly changed from a high-grain diet to a forage diet, generic *E. coli* populations in feces declined 1000-fold within 5 days, and the ability of the *E. coli* population to survive an acid shock similar to the human gastric stomach decreased (Callaway *et al.*, 2003). The suggested explanation was that when cattle are fed high-grain diets, some starch escapes ruminal microbial degradation and passes to the hindgut where it is fermented. Entero-hemorrhagic *E. coli* are capable of fermenting sugars released from starch breakdown in the colon, resulting in higher populations of *E. coli* in the gut and increased shedding of *E. coli* O157:H7. An all-forage diet reduced these populations in the gut by changing the pattern of digestion in the gut. Callaway *et al.* (2003) observed that other researchers have shown that a switch from a high-grain to a hay-based diet resulted in a smaller decrease in *E. coli* populations and without the effect on gastric shock survivability.

These findings suggest that, in general, the risk of *E. coli* O157:H7 contamination should not be greater with organic beef and is likely to be less, especially when concentrate feeding is stopped in the period before slaughter.

In The Netherlands, a study on 31 organic farms (Rodenburg *et al.*, 2004) showed a prevalence of 13 percent for *Salmonella* and 35 percent for *Campylobacter*. The incidence of *Salmonella* was lower and that of *Campylobacter* higher in organic than in conventional broiler flocks. In Denmark, *Campylobacter* was found to be significantly more prevalent in outdoor organic flocks than in both intensive and extensive conventional indoor flocks (Heuer *et al.*, 2001). In the USA, the prevalence of *Campylobacter* in organic and conventional retail chicken was similar (around 75 percent), whereas *Salmonella enterica* could be detected in a greater proportion of organic (61 percent) than conventional (44 percent) chickens (Cui *et al.*, 2005).

The incidence of three bacterial food-borne pathogens on organic and kosher poultry was reported by Nou *et al.* (2007). In total, 353 whole or cut-up raw poultry samples (104 conventional, 108 kosher, 41 kosher-organic, and 100 organic) from retail stores in Maryland and Virginia were tested for the presence of *Salmonella*, *Campylobacter* and *Listeria* over an 8-month period. *Salmonella*, *Campylobacter* and *Listeria* were isolated from 28, 49 and 45 percent of poultry samples, respectively. *Salmonella* was most frequently isolated from organic poultry samples (40 percent), as were *Campylobacter* from conventional (69 percent) and *Listeria* from kosher (67 percent) poultry.

Yersinia enterocolitica (another food-poisoning bacterium, often causing problems in people who eat chitterlings, i.e. pig intestines) was detected in 18 percent of organically raised pigs and in 29 percent of conventionally reared pigs (Nowak *et al.*, 2006).

In assessing the risks of food poisoning from meat it is fair to point out that food poisoning is more likely to be related to the way in which meat is handled after slaughter than its origin.

Cloning

Gardeners are familiar with cloning. As explained in Chapter 4, you can cut a shoot from a plant and stick it into a rootstock or a moist medium such as soil. After a short time the cutting will sprout roots and grow into a new plant, identical genetically to the parent. It is a clone. Sometimes the cutting has to be dipped in a rooting hormone powder to aid the rooting process. When it comes to animals, the process is not so easy and consumers have some apprehension about cloning animals.

The image of cloning was, for many people, made scary by the Gregory Peck film *The Boys from Brazil*. The film tells the fictional tale of ex-Nazi Dr Joseph Mengele's plan to produce 94 clones of Adolf Hitler. Having perfected his techniques by experimenting on twins in the Auschwitz concentration camp, Mengele sends out young baby boys to be adopted around the world. The boys are genetic clones of the dead Nazi leader. Mengele carefully selects the parents in an attempt to reproduce the family background and upbringing of the original Adolf Hitler. But Hitler's father had died when he was 13 years old, and Mengele needs to organize a team of former Nazis to kill all the adoptive fathers. The story was published in 1976 and made into a film in 1978.

The other event that brought cloning into the public spotlight was the birth of Dolly the lamb. In 1996 Dr Ian Wilmut and a research team at the Roslin Institute near Edinburgh in Scotland cloned the world's first sheep from adult cells, using the mammary cells of a ewe. The lamb was born on July 5 and was named Dolly after the country singer Dolly Parton. The procedure used by Wilmut's team was to implant genetic material from an udder cell of one sheep into the hollowed-out egg of another. That triggered cell division and the resultant embryo was implanted into a surrogate-mother sheep.

Since the birth of Dolly, scientists have cloned 10 other animals: cows, goats, pigs, rats, mice, rabbits, cats, dogs, horses and mules. A Korean company will clone your pet. The success of the technology has led animal geneticists to adopt it to produce superior breeding stock for use on farms and in artificial-insemination centers. Cattle are receiving most attention since the technology has the greatest potential to improve dairy and beef cattle. Traditional breeding methods with cattle take a long time because of the longer growth period to maturity and a low rate of reproduction. Pigs and poultry grow and reproduce fast, therefore there is less of an incentive to use cloning with these species.

The development is not without controversy. "It is playing God," say some. Other critics have suggested that the molecular machinery of the egg may be damaged during cloning. Support for this criticism was the fact that Dolly died prematurely, at least by breed standards. Most Finn-Dorset sheep live to be 11 to 12 years of age, but Dolly was put down by lethal injection in 2003 at just over 6 years of age. Prior to her death, Dolly had been suffering from lung cancer and crippling arthritis. Post-mortem examination of Dolly seemed to indicate that, other than the cancer and arthritis, she was quite normal.

The organic industry does not approve of animal cloning, therefore consumers who wish to avoid the possibility of eating meat from cloned animals should choose organic meat. It is unlikely, however, that meat from cloned animals will be available in the stores. The farming industry will use the technology to produce superior breeding stock which, like Dolly, will be used as the parent generation of superior offspring. It is the offspring of these matings that will be used for meat.

The US Food and Drug Administration has reviewed the situation and has ruled that products such as meat and milk that come from cloned animals are as safe to eat as foods from conventionally bred animals. In announcing its decision the agency stated:

> Following extensive review, the risk assessment did not identify any unique risks for human food from cattle, swine or goat clones, and concluded that there is sufficient information to determine that food from cattle, swine and goat clones is as safe to eat as that from their more conventionally bred counterparts.

The US Department of Agriculture supported the FDA final assessment. In its statement, the FDA explained that an animal clone is a genetic copy of a donor animal, and is similar to an identical twin. Cloning is not the same as genetic engineering, which involves altering, adding or deleting DNA. Because of the cost factors involved with cloning, such animals are intended for use as elite breeding animals to introduce desirable traits into herds more rapidly than would be possible using conventional breeding.

According to press reports in 2010 the meat from a few cloned animals in the UK (culled from breeding herds) inadvertently entered the food chain. In one case meat from the offspring of a cloned cow, raised and slaughtered in the UK, was exported to Belgium. However, Belgian authorities did not regard this as a food safety issue, and no rapid alert was issued. This was because the European Commission in 2010 informed all EU member states that, based on advice from the European Food Safety Authority, food from the progeny of clones are conventional foods and therefore no special measures apply.

Mad-cow disease

A horrifying disease affecting humans came into the limelight in the 1990s in Britain. It was eventually linked to the eating of beef products (specifically brain and related tissue) from cattle infected with "mad-cow disease". This was a variant form of the cattle disease. By 1996, several people in Britain and others who had lived there were affected with this form of the disease and the number rose to 143 by 2003.

"Mad-cow" is a disease affecting the brains of cattle. Technically the condition is known as bovine spongiform encephalopathy or BSE. It is one of a group of rare neurodegenerative disorders called transmissible spongiform encephalopathies (TSEs). Mad-cow disease is a chronic, degenerative disorder affecting the central nervous system of cattle. It is caused by rogue proteins, known as prions, that reproduce within the brain of the infected animal. As the disease progresses, the brain starts to look like a sponge, hence the term "spongiform". In addition to BSE, there is a similar condition called scrapie in sheep and Creutzfeld-Jakob disease (CJD) and kuru in humans. Chronic wasting disease (CWD), a related condition, has been reported in mule deer and elk in the United States.

These conditions cause neurological symptoms such as unsteadiness and involuntary movements that develop as the illness progresses, with late-stage sufferers being completely immobile at the time of death. Kuru, which was discovered in Papua New Guinea, was possibly transmitted by cannibalism when family members ate the body of a dead relative (including the brain) as a sign of mourning. The practice was banned in the 1950s, preventing further possible transmission. Other communities that like to eat brains have experienced similar illnesses. In 1997 a number of Kentuckians contracted a similar disease. It was discovered that all the victims had a liking for squirrel brains.

Research has shown that what causes these diseases is not a virus or bacterium as in other diseases, but a protein agent called a prion. The prion transforms normal proteins into infectious, deadly proteins that reproduce in the brain.

Other research has indicated that there is a genetic influence on the occurrence of the disease. Some people are more likely to develop it than others.

According to the US Centers for Disease Control and Prevention (CDC), 200 individuals worldwide were diagnosed with vCJD in 2006, including 164 people in the United Kingdom, 21 in France, four in the Republic of Ireland, three in the USA, two in The Netherlands, and one each in Canada, Italy, Japan, Portugal, Saudi Arabia, and Spain. Of these individuals, most (170) had lived in the UK for over 6 months during the years 1980–1996 and 20 others had lived in France during that time. The disease appeared therefore to be more prevalent in Europe, especially in the UK, than in other regions.

Kugler (2006) summarized the relevant facts about mad-cow disease and humans. She explained that the public has good reason to be concerned about the possible transmission of BSE to humans. Variant Creutzfeldt-Jakob disease, like the other types of Creutzfeldt-Jakob disease, is a brain disorder that becomes deadly over time. There is no cure.

Normally, Creutzfeld-Jakob disease occurs in a person in one of three ways:

(1) About 10–15 percent of cases are inherited, resulting from a gene mutation.
(2) Most cases seem to appear sporadically in someone who has no family history of the disease.
(3) A small percentage of cases occur through infection, by contact with infected brain tissue.

According to Kugler (2006) there are documented cases that have occurred as an unintended consequence of a medical procedure.

Creutzfeldt-Jakob disease is not contagious in normal ways, like sneezing or coughing, and there are no known cases of spouses or family members of an infected person contracting the disease. All prion diseases are fatal. There is no effective treatment.

Contaminated beef products implicated

The outbreak of variant Creutzfeldt-Jakob disease in the UK in the 1990s seems to have been caused by people eating contaminated beef products. In the mid-1980s an epidemic of BSE was seen among cattle in the UK. The suspected cause was a prion transmitted in meat and bone meal products fed to cattle, therefore the government banned the practice of feeding these products to cattle in 1988. Other countries did likewise. By then, however, infected cattle had already entered the human food supply in the United Kingdom.

Most countries have now instituted regulations to prevent mad-cow disease from spreading to consumers. First, older cattle are not allowed to enter the human food supply and a health check is made of animals sent for slaughter. Secondly, all high-risk beef tissues such as brain and spinal cord tissue are removed when the animals are slaughtered and are not allowed to enter the human food supply. Thirdly, the feed regulations ensure that the feed contains no animal products (except milk products). Veterinary regulations have been enacted, including restrictions on the importation of breeding animals, to eradicate mad-cow disease from beef herds. For instance, Australia is now categorized as a country free of indigenous BSE, by the use of measures such as these.

The regulations have worked well in North America in protecting the public from being exposed to mad-cow disease. Also, it is important to emphasize that there was never any problem associated with beef cuts such as steak or roasts.

Although Britain was the country that experienced most cases of mad-cow disease, it is interesting to note that this issue is no longer identified as a major concern by the British consumer.

There are instances of public alarm in some countries as evidenced by refusal to accept North American beef because of mad-cow disease problems. For example, there was uproar in South Korea over American beef in 2008, with riot police called to quell protests by tens of thousands of people. Six months later the *Washington Post* reported, "What was the big deal? Low-priced US beef has appeared in supermarkets here in recent days, after a decision by three major retailers to start selling it again, and the reaction has been brisk business and no political fuss. Fifty tons of US beef disappeared from shelves the first day it was offered for sale."

So was the Korean ban due to concern for human health or protectionism for their own farmers? The answer seems to be that, as with many commodities, countries take steps to protect their own producers and will use isolated disease outbreaks to justify an import ban. World Trade negotiations are frequently involved in situations such as this.

Gene modification

Gene modification is different from cloning. In cloning, the whole embryo is removed from one cell and inserted into another. In gene modification, the molecular structure of the embryo is altered.

Work on the commercial application of gene modification of animals is currently less active than with cloning. Among the research projects underway is a project at the Australian Animal Health Laboratory to breed a chicken that is resistant to avian influenza. The virulent strain of the causal virus – the H5Nl strain – has led to 324 reported human deaths (as of June 3, 2011) in 15 countries and the culling of thousands of birds. The aim of the research is to create resistance to viral infection which is then integrated into the DNA so that it can be passed on to future generations of chickens through normal breeding.

The US Council for Agricultural Science and Technology recently reviewed the scientific evidence related to the safety of meat, milk and eggs from animals fed crops derived from modern biotechnology (CAST, 2006). Their review concluded that:

Performance, health, and nutrient use by farm animals are similar when fed either conventional or biotechnology-derived crops, and/or their co-products. Furthermore, no biologically relevant differences in the composition of animal products, including milk, meat and eggs, have been reported between farm animals fed diets containing commercially available, biotechnology-derived crops and those fed diets containing conventional genetic counterparts. No intact or immunologically reactive fragments of transgenic plant proteins or deoxyribonucleic acid (DNA) have been detected in samples of meat, milk, eggs, lymphocytes, blood, and organ tissue from production animals fed biotechnology-derived crops modified for agronomic input traits. The regulatory processes in place to assess the safety of biotechnology-derived crops have been effective in safeguarding public health. To date, there has been no authenticated case of an adverse health-related incident associated with the consumption of food or feed derived from modern biotechnology. The review of the currently available data concludes that meat, milk and eggs produced by farm animals fed biotechnology-derived crops are as wholesome, safe and nutritious as similar products derived from animals fed conventional crops.

An international panel of scientists reviewed the report and concurred with the findings. This report can be regarded, therefore, as an important reassurance that public safety has not been compromised by the introduction of biotechnology-derived foods.

Further reassurance on this issue was provided in a scientific report presented at the BIO 2008 International Convention (Gottlieb and Wheeler, 2008). The authors argued that genetically engineered animals embody an innovative technology that is transforming public health through biomedical, food and environmental applications. Other perceived benefits were as follows:

- They are integral to the development of new diagnostic techniques and drugs for human disease, while delivering clinical and economic benefits that cannot be achieved with any other approach.
- They promise significant benefits in human health and food security by enabling dietary improvements through more nutritious and healthy meat and milk.
- Genetically engineered animals also offer significant human health and environmental benefits with livestock which are more efficient at converting feed to animal protein, and reducing waste production.
- Genetic engineering will improve the welfare of the animal by imparting resistance to disease and enhancing overall health and well being.

The US Food and Drug Administration has announced details of how it will approve the entry of genetically altered animals into the food supply (FDA, 2009). These will be regulated in the same way that drugs for animals that enter the food supply are regulated. Companies that wish to bring genetically altered meat to the market would have to prove that the alteration does not pose any danger to the health of the animal or the consumer.

Meat that has altered nutritional content would require to be labeled to reflect the change. Otherwise, genetically altered meat would require no special labeling. That requirement is similar to the way in which food from genetically altered crops such as corn and soybeans is labeled.

Nutritional and organoleptic qualities

Beef

Research findings

Beef quality is assessed in two main ways: grade and eating quality. The payment grade of beef carcasses is determined primarily by estimates of lean meat and fat contents. For instance, beef graded as Prime by the US Department of Agriculture (about 2 percent of graded beef) has more marbling (fat within the meat). However, it is higher in fat content. The qualitative traits of beef, such as color, tenderness and flavor, are not taken into account in setting the payment grade. Inspection of slaughtered animals in the USA is mandatory, but grading is voluntary and a plant pays to have its meat graded. The payment grade achieved by organic beef carcasses often underestimates their commercial value because they contain a lower level of fat than conventionally raised beef animals.

A main difference between the organic and conventional beef production systems is that the organic animals are fed mainly or exclusively on grass and other forages. In conventional production cereal grains are also included in the feed. This difference results in the animals fed organically growing more slowly and taking longer to reach market weight. Also, the carcasses are leaner. The longer time needed to reach market weight helps to explain the higher cost of organic beef to the consumer.

Russo and Preziuso (2005) reviewed findings in the scientific literature on the qualitative characteristics of carcasses and beef from cattle raised organically. According to their findings, carcasses of organically raised beef cattle are characterized by poor muscular development and reduced fat content. They attributed this to the fact that the diets are mainly based on forage, with low energy contributions from concentrates. Native breeds are preferred in organic production, especially in Europe, and are often characterized by rather slow development. These facts explained a common finding that organic beef has a lower intramuscular fat content than beef from cattle raised conventionally. Another conclusion was that other organoleptic characteristics of the meat do not appear to be influenced by the organic rearing system.

Table 6.10   **Effect of grass feeding on the carcass characteristics and meat quality of steers**

	Feeding system		
	Grass alone (18 kg dry wt)	Grass + concentrate (12 kg grass dry wt + 2.5 kg concentrate	Concentrate to appetite, no grass (plus 1 kg straw)
Carcass weight (kg)	330	348	371
Carcass gain/day (g)	360	551	809
Fat score[1]	4.03	4.15	4.64
Kidney fat[2] per carcass (g/kg)	24	22	29
Intramuscular fat (marbling, g/kg muscle)	23	25	44
Moisture content (g/kg muscle)	737	734	717
Protein (g/kg muscle)	225	228	226
Meat quality after 2 days of aging			
Warner–Bratzler shear force	8.0	6.4	6.1
% cooking loss	30.0	29.1	29.8
Tenderness[3]	3.5	4.8	4.4
Texture[4]	2.9	3.2	3.3
Flavor[5]	3.5	3.6	3.7
Juiciness[6]	4.8	4.7	5.2
Chewiness[7]	4.2	3.6	3.9
Acceptability[8]	3.2	3.2	3.3

[1] 1=Leanest–5=fattest.
[2] Kidney plus channel fat.
[3] 1=Extremely tough–8=extremely tender.
[4] 1=Very poor–6=very good.
[5] 1=Very poor–6=very good.
[6] 1=Extremely dry–8=extremely juicy.
[7] 1=Not chewy–6=extremely chewy.
[8] 1=Not acceptable–6=extremely acceptable.

(*Source*: French, P., O'Riordan, E.G., Monahan, F.J., *et al.* (2001). The eating quality of meat of steers fed grass and/or concentrates. *Meat Science* **57**, 379–386. © Elsevier Science. All rights reserved.)

These effects can be seen in results obtained by French *et al.* (2001). Results for the extremes of treatments used, i.e. grass alone or concentrate feed alone (plus some straw) with no grass, and for a more typical feeding system (grass plus concentrate) are shown in Table 6.10. Animals were slaughtered after an average of 95 days. Samples of rib beef were collected after 2, 7 or 14 days of aging. Taste assessment by a trained panel and other tests of the quality of the beef were then conducted.

As reported by other researchers, animals fed grass alone take longer to reach market weight. Fat score and intramuscular fat content were signifi-cantly higher in the animals fed all-concentrate, and were intermediate in the animals fed grass plus concentrate. Muscle moisture content was also significantly lower in the animals fed the all-concentrate diet. There was an obvious trend for beef quality as judged by a taste panel and by laboratory testing to improve with an increasing level of concentrate in the diets. However, the differences failed to achieve statistical significance. A factor having a greater effect on meat quality was the period of aging after slaughter.

It is known that grass-fed beef can be of lower eating quality than grain-fed beef. For instance, Woodward and Fernández (1999) found that beef animals fed a conventional finishing diet had heavier carcasses and larger rib-eye muscles than animals fed organically, but the meat had less marbling. Melton *et al.* (1982) found that ground beef from animals finished on grass was lacking in beef fat flavor, had a more intense dairy-milky flavor and usually had a soured dairy or other off-flavor. Muir *et al.* (1998) concluded that the differences in flavor and acceptability due to feed type can be accounted for by differences in carcass fatness. The lower quality reported in some studies might be due to the use of poor-quality grasses and forages, resulting in slow growth and a low fat content in the carcasses of beef cattle fed organically.

In a review of organic beef production in Denmark, Nielsen and Thamsborg (2005) found that grazing and increased animal activity, which are inherent components of an organic beef production system, may affect the eating quality due to darker meat color, risk of off-flavor, yellow fat, and a higher content of unsaturated fatty acids, including conjugated linoleic acid (CLA). They concluded, however, that the overall effect on sensory attributes may be of minor importance.

An Italian study (Miotello *et al.*, 2009) compared the quality of veal from beef cattle raised organically or conventionally. The contents of fat and cholesterol were significantly lower in organic veal than in conventional veal. Cooking loss was also lower in the organic veal. The red color index was higher in organic veal, due to a higher content of heminic iron. The tissue fat in organic veal was found to have a more desirable profile of *n*-3 fatty acids, *n*-6/*n*-3 ratio and CLA content than in conventional veal.

As indicated earlier, the amount of marbling in the meat affects beef quality and flavor. This feature is influenced by breed as well as feed. For instance, Dinh *et al.* (2010) reported significantly greater concentrations of saturated fatty acids (26.67 mg/g), monounsaturated fatty acids (26.50 mg/g), and polyunsaturated fatty acids (2.37 mg/g) in longissimus muscle in Angus cattle than in Brahman or Romosinuano cattle. As a result, a distinct trend in organic beef production, especially in Europe, is the use of heritage breeds that have higher marbling in the meat. In Asia it has been shown that breeds such as Japanese Black cattle have a genetic predisposition for producing carcass fat containing higher concentrations of monounsaturated fatty acids than Holstein, Japanese Brown or Charolais cattle. Also, Welsh Black cattle (a traditional beef breed) deposit higher proportions of C18:3n-3 and its metabolic products C20:5n-3 and C22:5n-3 in muscle phospholipids and higher proportions of C18:3n-3 in muscle neutral lipids and adipose tissue than Holsteins even after the dietary intakes of n-3 PUFA were increased by feeding linseed or fish oil. These findings indicate a genetic variability in fatty acid synthesis and deposition among breeds that influences both marbling and its lipid composition. The appropriate choice of breed is therefore becoming important for organic producers seeking to achieve a desired content of marbling in the meat.

A main effect of forage feeding on beef composition is to alter the fatty acid composition of the carcass fat, influencing both the nutritive value of beef and also the organoleptic properties, in particular flavor. Dietary effects on beef composition are relatively well recognized and in general the composition of the beef lipids reflects the fatty acid composition of the diet. For instance, beef can be a relatively rich source of n-3 PUFAs due to the presence of C18:3 in grass. Razminowicz *et al.* (2006) reported that the year-round feeding of forage products results in n-3-enriched beef, which is at least as tender as conventional beef.

Currently there is considerable interest in developing production methods aimed at raising the levels of these fatty acids in beef to improve meat quality. Such systems provide a product which is regarded as "healthier" and more attractive to the consumer and which, in turn, may command a price premium.

Conjugated linoleic acid (CLA) consists of a group of isomers of linoleic acid to which anti-cancerogenic, anti-diabetic, and anti-atherogenic effects, as well as effects on the immune system, bone metabolism and body composition, have been attributed. CLA is found predominantly in milk and meat of ruminant animals due to the action of rumen microorganisms in the formation of CLA and its precursors.

A switch from a concentrate-based diet to pasture has been shown to increase the CLA content of beef. For instance, French *et al.* (2000) reported increasing CLA contents in the intramuscular fat of steers (longissimus dorsi muscle) that were consistent with increasing intakes of grass. Levels of 5.4, 6.6 and 10.8 mg CLA/g were detected in grazing steers with increasing grass intake compared to 3.7 mg/g in animals fed concentrate. Grass silage also positively influenced CLA content (4.7 g/g) but not to the same extent. Poulson *et al.* (2004) reported a 6.6 times higher CLA content in the longissimus and semitendinosus muscle from steers raised only on forages compared with steers fed a common high grain feedlot diet (13.1 vs. 2.0 mg/g).

A concern is that a high content of PUFA in meat might result in a shorter shelf-life (due to lipid and myoglobin oxidation) and reduced flavor because of the instability of these fatty acids. However, it appears that only when concentrations of α-linolenic acid (18:3) approach 3 percent of neutral lipids or phospholipids are there any adverse effects on meat quality (Wood *et al.*, 2003). In addition, grazing provides antioxidants including vitamin E which maintain PUFA levels in meat and prevent quality deterioration during processing and display.

Consumer findings

Overall, most studies report that consumers purchase organic foods because of a perception that such products are safer, healthier and more environmentally friendly than conventionally produced alternatives. Also, consumers expect substantially higher quality in meat produced in organic and pasture-based systems, which are perceived as being more "natural".

The production of organic beef continues to increase in response to consumer demand even though there is evidence of a substantial variation in the quality of organic meat entering the marketplace (e.g. Sundrum, 2010), also that organic meat costs more at the retail level than conventional meat, due mainly to a more restricted supply of acceptable feedstuffs.

The purchase of meat by the consumer appears to be governed by two main factors:

(1)   The initial perception and expectation of quality based on appearance, price, presentation and labeling, and possibly ethical and philosophical considerations such as freedom from chemical residues and how the animal was raised.
(2)   The actual quality experienced after cooking and eating.

The response to the second factor greatly influences whether the consumer purchases the same meat on other occasions. European research indicates that the first factor is much more important than the second (e.g. Scholderer *et al.*, 2004).

Grunert (2006) found that once a perception is firmly established in the mind of the consumer the effects on quality perception can be quite dramatic. Both country of origin and organic production have been shown to have "halo effects" with regard to quality perception. Consumers tend then to believe that organic meat is better not only in terms of how it was produced, but also in terms of "healthiness" and sensory quality. Also, in situations where the physical differences between alternative products are small, the quality inferences made on the basis of these cues may be so strong that consumers stay with their choice regardless of other information.

Although there is a strong demand for organic meat, there is evidence that it has to be produced economically. A survey in Scotland found that organic meat was perceived as being expensive, especially when consumers did not perceive a positive difference in quality (McEachern and Schröder, 2002; Andersen *et al.*, 2005). This resulted in some consumers being more interested in conventional meats with added-value features (e.g. produced to high standards of animal welfare) rather than organic meat. A Canadian study confirmed the price aspect. Anders and Moeser (2008) estimated the consumer demand for organic and conventional fresh beef products in the Canadian retail market and found that demand for organic beef was highly dependent on price and expenditures.

Several consumer studies have found regional differences in how consumers perceive meat quality. For instance, Corcoran *et al.* (2001) reported on consumer attitudes towards lamb and beef in Europe. Factors that characterized meat quality were largely similar in all focus groups. Color was first, though color preference differed between countries. In Scotland bright red was considered false, suggesting additives or a lack of maturation. A natural red was preferred for beef. In Spain, intense red or brown was not appreciated and consumers preferred a "pinkish" color. In Italy an intense red color was preferred, but not brown. The latter was not considered fresh. A strong odor was disliked everywhere. British and French participants favored a marbled appearance in beef for enhanced flavor. The Spanish preferred fresh rather than mature meat. They also refused "cheap" meat and considered origin to be a very important quality attribute. Italian participants cared more about freshness, a low fat content and type of packaging. Ready-frozen meat was not appreciated by any of the participants in the five locations. Some UK participants were confused, ignorant and/or mistrusting of the number of current assurance schemes covering meat quality. They felt that an independent body or consumer group, not government, should control meat safety and quality. The Spanish felt that the government was responsible for guaranteeing safety and quality but they thought that farmers should also be responsible for delivering these. Italian participants put their trust in the public health services. All participants said they would pay more for a better level of assurance, quality and information relating to the beef and lamb meat and meat products that they purchased. In general, participants complained about a lack of clear and consistent information on quality meat products. More information was sought by all participants: the British wanted information on "eatability", production methods, and animal welfare; the Spanish, days to market, farmer and origin; the Italians, more nutritional information and information on origin; and the French, information on production and processing, details of health implications and control of "taste" quality.

As indicated in a previous section, one concern that causes some consumers to hesitate about buying organic meat from animals raised outdoors is the possibility of bacterial contamination and the risk of food poisoning.

The survey responses indicate large regional differences in consumer attitudes and taste, which should be taken into account in producing organic meat for the region in question.

There is also evidence that nutritional quality is becoming at least as important to the consumer of organic beef as safety concerns. A factor related to this is an altered profile of the fat, which is regarded as being more favorable in terms of human health. This altered profile, with a higher content of polyunsaturated fatty acids (PUFAs), should theoretically make organic beef more susceptible to rancidity and to have a lower shelf-life than conventional beef, but to date there is no evidence of any problem in this regard.

Pork

Research findings

Organic feeding has less of an effect on pork quality than on beef quality, because the diet is not changed as dramatically. Although organic pigs are allowed to graze, their digestive systems are not designed to handle a high intake of a fibrous feed such as grass. Consequently their diet is not too dissimilar to that fed to conventional pigs. The main difference is that only organic sources of feed grains and protein supplements are used with organic pigs (Blair, 2007). Differences in pork quality have nevertheless been reported.

The results of Sundrum *et al.* (2000) in Germany typify how organic diets affect the growth and carcass quality of pigs. A conventional feed mixture was compared with three feed mixtures containing different organic protein sources that could be grown on-farm (faba beans + potato protein; peas + lupinseed; or faba beans + lupinseed). These protein sources replaced solvent-extracted soybean meal which is not permitted in organic feed mixtures.

Pigs fed the organic diets ate less and grew more slowly than pigs fed the conventional diet. This was attributed to a better balance of amino acids in the conventional diet due to the inclusion of pure lysine. As a result they were more costly to produce. Carcass quality differed also when the pigs were slaughtered at a similar weight (Table 6.11).

The percentage of lean meat and the size of loin muscle were lower, and the amount of fat and the intramuscular fat were higher in the pigs fed organic diets. No other measures of meat quality were reported in the study, such as flavor or tenderness.

A large-scale investigation conducted by Danish Pig Production and the Danish Meat Research Institute (Søltoft-Jensen, 2010) showed that organic feed affected tenderness as well as the color and taste of pork chops in comparison with a conventional diet. Two organic diets were tested, one with 80 percent inclusion of organic ingredients and the other 100 percent. The reason for the two organic diets was that the European organic regulations currently allow some non-organic feedstuffs to be used because of a scarcity of organic ingredients.

Table 6.11 Carcass quality of pigs fed conventional or organic diets (Sundrum *et al.*, 2000)

Carcass parameter	Diet type			
	Conventional	Faba beans + potato protein	Peas + lupinseed	Faba beans + lupinseed
Slaughter weight, kg	93.1	92.1	91.2	91.7
Carcass yield, %	77.9^e	76.9^{ef}	76.7^{ef}	76.5^f
Lean meat, %	56.0^e	55.6^{ef}	54.3^{fg}	53.6^g
Longissimus area, cm^2	56.8^e	54.3^e	48.8^f	48.0^f
Lean:fat ratio.	0.33^e	0.33^e	0.39^f	0.39^f
Fat area, cm^2	18.4	18.0	19.0	18.6
Backfat thickness, cm	2.4	2.4	2.4	2.4
Intramuscular fat, %	1.20^e	1.25^e	2.90^f	2.95^f

e,f,g Values within a row with different letters differ ($P < 0.05$).

Chops from pigs fed 100 percent organic feed were less tender and the meat color was slightly darker than in chops from pigs fed a conventional diet. This was attributed to the slower growth of the organic pigs. This result confirms previous findings of an improvement in tenderness and eating quality of pig meat from animals with a high growth rate.

The Danish researchers found that the conventional pork chops were more tender and less tough, and appeared more crumbly and less crisp and stringy than chops from pigs fed organically. Chops from the 100 percent organically fed pigs differed from chops from the 80 percent organically fed pigs by being less tender and having a greater bite resistance/toughness and by being less crumbly. The difference between chops from 100 percent organically fed and conventionally fed pigs was as much as 3.7 units, a difference that would be detected by the ordinary consumer. Usually a difference of 1 unit can be detected by a taste panel. The chops also had a different taste. Those from pigs fed the 100 percent organic diet were more "piggy" and had more of a metallic odor than chops from pigs fed the 80 percent organic or conventional diet. Color measurements showed that meat from pigs fed the conventional diet was lighter than the meat from pigs fed the two organic diets. The difference in color between meat from conventionally fed pigs and the 80 percent organic feed pigs would not be noticed by the ordinary consumer. However, meat from pigs fed 100 percent organic feed was considerably darker that the meat from the other two groups of pigs and would be noticed by the consumer.

The researchers suggested that the differences in meat tenderness and meat color were most likely related to differences in the growth rate of the pigs. Research in Denmark and in other countries has shown that a large gain in muscle development prior to slaughter results in an increase in meat tenderness. Investigations have also shown that a reduced rate of weight gain affects the color of the meat.

These results pose somewhat of a dilemma for European organic pig producers. The feed has to be 100 percent organic by 2012.

The effects on fat content reported by Sundrum *et al.* (2000) (see Table 6.11) are considered desirable by some lovers of pork who feel that production of lean carcasses to meet a consumer demand for low-fat meat has resulted in a substantial decrease in intramuscular fat levels and a decrease in eating quality. The desire for higher levels of marbling in pork is reflected also in an increased use of heritage breeds in organic pork production, similar to the trend in organic beef production. Mediterranean silvopastoral systems have been associated with improved quality of pig meat when used for dry-cured ham production, due to the relatively high levels of intramuscular fat resulting from access of native pig breeds to acorns which are high in starch (Edwards and Casabianca, 1997).

In addition to these findings that diet can have an important effect on the amount of fat in the pig carcass, it has also been shown that the fatty acid composition of the fat can be altered with organic feeding to produce a higher content of PUFA (Blair, 2007). This has important implications in terms of the keeping quality of the meat and its suitability for further processing.

The suggestion has been made that pork fat with an enriched content of PUFA is more desirable for the consumer in that it would help to promote better health. This is relatively easy to achieve as PUFA are readily incorporated in pork fat. Studies have demonstrated that this can be achieved by the use of feedstuffs such as canola seed, linseed or soybeans, and the changes in fat composition can be obtained in a relatively short time. Also, pigs given access to pasture have higher levels of PUFA, n-3 fatty acids and vitamin E in the muscle than indoor pigs. Pork fat enhanced in this way could be a selling feature for organic pork. However, it would be important to ensure that the animal had a high intake of vitamin E to prevent oxidative changes in the fat, and to market the meat soon after production to avoid rancidity of the fat. Such meat would be best used fresh and not stored or used for further processing. High amounts of PUFA in the feed can have deleterious effects on meat softness, meat storage stability and quality (Jakobsen, 1999).

The presence of PUFA in the fat of pigs indicates their presence in the diet, since pigs fed a diet containing no fat or a low level of fat synthesize and deposit saturated (mainly 16:0 and 18:0) and monounsaturated (18:1) fatty acids.

Consumer findings

As with beef, the purchase of pork meat by the consumer appears to be governed by two main factors:

(1) The perceived quality based on appearance, price, presentation and labeling, and ethical and philosophical considerations such as freedom from chemical residues, etc.
(2) The actual quality experienced after cooking and eating.

European research indicates that, as with beef, the first factor is much more important than the second. Consumers expect substantially higher quality in pork produced in organic and free-range systems. The results obtained by Scholderer *et al.* (2004) suggest that the actual or experienced quality of organic pork is very much a matter of expectation. The authors concluded that consumers appeared able to detect the lower eating quality of organic pork, but that the difference between the high expectations and moderate experiences of the quality of the organic pork appeared to fall within the range of acceptance. Had the discrepancy between expectation and experience been higher, it might have resulted in rejection of the organic pork.

One factor which could have influenced the results was that, contrary to what one might have expected, the loin meat of the organic pigs was slightly wetter (74.9 vs. 74.5 percent) and contained less intramuscular fat than the meat from the conventional pigs (1.8 vs. 2.0 percent). This difference in meat composition may explain the ratings reported by a taste panel.

In other research in this area Bredahl and Poulsen (2002) reported results from a focus group study with Danish pork consumers. In this study, consumers defined pork high quality in terms of good taste (most important), tenderness, juiciness, freshness, leanness and "healthiness". Participants generally found it difficult to evaluate the quality of meat at the point of purchase, particularly pre-packaged meat. Color of the meat and fat content were important quality attributes. Generally the participants preferred meat with little fat cover. To some extent color of the fat and of the meat were used as indicators of freshness. In addition, the results showed a clear perceptual link between the quality of pork and the production method, in which extensive outdoor production was generally perceived to result in higher quality than intensive indoor production. This attitude was described by Scholderer (2003) as the "halo effect". Factors relating to quality included transportation of the animals, how they were kept at the farm, what feed was used, use of growth enhancers, treatment of the live pigs at the slaughterhouse, general welfare of the animals, use of medicine, breed of pigs, level of veterinary control, and the cooling of the meat. Despite the preference for "welfare" pork, meat from extensive production systems was rarely bought. Although it was generally regarded as desirable, the participants generally rejected the meat because they perceived it to be either too expensive or too difficult to obtain.

Results of these studies showed that Danish consumers believe that good rearing conditions, natural feeding of the pig, non-use of antibiotics, leaner meat and generally higher quality are important attributes of organic pork, and they associated these attributes with physical well-being, good health and a long life. There appeared to be at least four different dimensions of considerations when choosing or not choosing organic pork: animal welfare: higher organoleptic quality of organic pork, expectations regarding health, and budget constraints. Budget constraints, more than availability, appeared to be the most important barrier to increased purchase and consumption of organic pork (Bredahl and Poulsen, 2002).

These findings probably apply to most consumers.

Diet does not appear to have a major influence on pork flavor, unless a feedstuff such as fish meal is included (Melton, 1990). It can result in a fishy flavor in pork when included at high levels in the diet. The effect on pork flavor is reduced or eliminated if the fishmeal is removed from the diet at least 2 weeks prior to slaughter. It was found that a sensory panel could not detect any difference in the flavor of country-cured hams from pigs fed primarily corn and those fed peanuts in the hull, even though the ham fat produced by pigs fed the peanut diet was much softer than in the pigs fed corn. The effects of different dietary ratios of barley and roasted soybeans were compared with those of maize and soybean meal on the flavor of cooked pork. No significant dietary effect on flavor of the pork was found, and the flavor of all of the pork samples was rated by a trained sensory panel as "liked moderately" to "liked very much". Related work showed that source of protein did not affect flavor score, but pork from pigs fed a milo-based diet had a less desirable flavor than pork from pigs fed corn, barley or wheat, although it was liked moderately. Other findings suggest that several protein supplements can replace soybean meal in the diet without effect on meat flavor. An exception is the use of high-oil protein feeds such as whole soybeans, due to the unstable nature of the oil which can become oxidized and result in off-flavors unless the diet is supplemented with vitamin E. Feeding such diets has been shown to increase the degree of unsaturation of the backfat due to an increase in the concentration of oleic (18:1) and linoleic (18:2) fatty acids and a corresponding decrease in palmitic (16:0) and stearic (18:0) acids.

Melton (1990) suggested that, perhaps because pork fat is already unsaturated compared with fat in other red meats, oxidative rancidity is part of an acceptable or intense pork flavor. A study by Cameron and Enser (1991) suggested that the presence of unsaturated fat in the diet might influence the eating quality of the pork. They found that correlations between PUFA and palatability scores were generally negative and those between the saturated fatty acids and palatability were generally positive. These results suggest that, the higher the degree of unsaturation of the intramuscular fat, the lower the palatability score and the greater the incidence of abnormal flavors. Another study involved three different combinations of beef fat and soybean oil in the diet to produce backfat containing 10 percent (normal), 20 percent (medium) or 30 percent (high) concentrations of 18:2. It was found that a consumer panel preferred the taste of back bacon made from pigs fed the normal or the high 18:2 diet. However, they preferred the taste of pork loin roll from pigs fed the diet with the medium 18:2 level over that from pigs fed the diet with the normal level of 18:2.

The above findings confirm that diet can have an important effect on the amount and fatty acid composition of fat in the pig carcass. This has important implications for the organic producer, mainly in terms of the keeping quality of the meat and its suitability for further processing. However, dietary composition appears to have minimal or no effects on the flavor of the meat itself. It would be important to ensure that the animal had a high intake of vitamin E to prevent oxidative changes in the fat and to market the meat soon after production to avoid rancidity of the fat. Such meat would be best used fresh and not stored or used for further processing, to avoid the problem of "warmed-over flavor" during storage and processing of the organic meat. Therefore, the quality of organically produced meat can be a matter of concern, especially its fatness and palatability.

Organic pigs are allowed access to grazing which can have an effect on pork quality. Animals fed unrestricted amounts of regular feed have been found to have very low forage intakes (<5 percent of daily DM intake; Blair, 2007). However, when the regular feed is restricted, the intake of forage can be up to 15 percent. Forage contains unsaturated fatty acids and antioxidants, and it has been shown that pigs given access to pasture have higher levels of PUFA, n–3 fatty acids and vitamin E in the meat than indoor pigs. This effect is considered desirable, but a less desirable effect is that a high intake of roughage can result in a reduced intake of nutrients and a lowered growth rate. As indicated previously, research has consistently shown an improvement in tenderness and eating quality of pig meat from animals with a high growth rate. Several researchers have reported that a reduced level of concentrate feeding and a high intake of forage (either clover grass or clover grass silage) results in a reduction in growth rate and reduced intramuscular fat content and tenderness of the meat (e.g. Hansen *et al.*, 2001).

Mediterranean silvopastoral systems have been associated with improved quality of pig meat when used for dry-cured ham production, due to the relatively high levels of intramuscular fat resulting from access of native pig breeds to acorns. Cava *et al.* (1999) showed increased aroma and flavor, and less rancidity in dry-cured ham, by feeding acorns to pigs.

The above studies suggest several important conclusions. First, organic pork should be produced in such a way that it lives up to the expectations of consumers. Secondly, the willingness of consumers to pay a premium for organic pork is not unlimited. These conclusions indicate that organic producers need to strive to produce a high-quality product as economically as possible.

Poultry

Research findings

As with pigs, organic feeding has less of an effect on poultry meat quality than on beef quality because the diet is not changed as dramatically. The diets fed to poultry and pigs are fairly similar though the cereal grains do not have to be ground for poultry. They possess a gizzard, which is an effective grinding organ. Although organic poultry are allowed to graze, their digestive systems, like those of pigs, are not designed to handle a high intake of a fibrous feed such as grass. Consequently their diet is not too dissimilar from that fed to conventional poultry. The main difference is that only organic sources of feed grains and protein supplements are used with organic poultry (Blair, 2008). Differences in meat quality have nevertheless been reported.

Jahan *et al.* (2005) compared the sensory qualities of organic, free-range and corn-fed chicken breast meat purchased in retail stores and super-markets, using a trained taste panel. In terms of appearance the organic meat was less highly pigmented and was more moist. No differences in aroma were detected, except that the organic meat was more "oily" and there was a slightly higher incidence of "sour" aroma. Meat from corn-fed birds was found to have a more intense "chicken" flavor, also to be more "buttery" and "salty". The texture of the organic meat differed also. It was found to be tougher, less succulent, softer and with more of a "cardboard" texture than the corn-fed chicken.

Other aspects of quality were investigated by Husak *et al.* (2008), also involving a comparison of broiler chickens purchased commercially as organic, free-range or conventional. The raw meat yield, cooked meat yield, proximate composition, pH, color, lipid oxidation score, fatty acid compo-sition, and sensory attributes were measured. Organic broilers had a higher yield of dark (thigh) meat than free-range or conventional broilers, when compared on a raw-meat basis. Conventional and free-range broilers yielded more cooked breast meat than organic broilers. The pH of breast meat from organic broilers was higher than in free-range or conventional chickens. Organic breast and thigh meat was less yellow than free-range or conventional. Fatty acid analysis showed that organic breasts and thighs were lower in saturated and monounsaturated fatty acids and higher in polyunsaturated fatty acids than free-range and conventional broilers. Shear force measurements were less for both breast and thigh meat from conventional broilers relative to free-range and organic broilers, indicating increased toughness in the meat from free-range and organic broilers. Sensory panel results confirmed that thighs from conventional broilers were more tender and less chewy than thighs from free-range and organic broilers, whereas other sensory properties did not differ.

Brown *et al.* (2008) raised broiler chickens under conventional, corn-fed, free-range or organic conditions and compared the effects on meat quality and sensory characteristics. Significant differences were found between rearing systems, with fillet muscles from birds grown under the conven-tional system having a higher ultimate pH. Differences were also seen in color with fillets from birds reared conventionally having a smaller hue angle than those grown using the corn-fed system which had the highest. Fillets from birds reared in the conventional system were rated by the taste panel as more tender and juicy. There were no significant differences in chicken flavor. Based on taste panel assessment, meat from birds produced in the standard system was most preferred and that from organic systems the least preferred. Meat from free-range and corn-fed systems was intermediate in preference and acceptance.

Somewhat similar results were obtained by other researchers. Symeon *et al.* (2009) studied the quality of market chickens and found that organic poultry meat had a lower content of intramuscular fat and a higher cooking loss. There were no significant differences for total moisture, crude protein and Warner-Bratzler shear values. Conventional chicken breasts scored significantly higher for juiciness and tenderness, but no significant difference was found for color, taste and overall acceptance. Grashorn and Serini (2006) compared the meat quality of fresh organic broilers purchased directly from farms and fresh conventional broilers purchased from supermarkets. Broiler weight was between 1300 and 3300 g for organic chickens and between 1100 and 1500 g for conventional chickens. Proportion of breast meat was found to be lower in organic chickens, the skin and the meat were more yellow, grilling losses were lower and texture values were higher. Organic chicken meat was drier and contained higher levels of protein, ash, fat and n-3 fatty acids. Sensory panelists assessed the organic broiler meat as tougher but tastier.

On the other hand, Castellini *et al.* (2002) compared the carcass and meat quality of the same breed and strain of hybrid broiler chickens raised conventionally indoors or organically indoors with access to a grass paddock and obtained better results for the organic birds. Results showed that carcass yield was significantly different between the two management systems in that organically produced carcasses had a higher proportion of breast and drumstick meat and a lower level of abdominal fat. In terms of meat quality, the organic chickens had lower water-holding capacity, increased cooking loss and increased muscle shear value (indicating increased toughness). A sensory panel ranked organic chicken higher in terms of juiciness and overall acceptability. The organic chickens had a higher proportion of saturated and reduced levels of monounsaturated fatty acids in the meat. In addition they had higher levels of polyunsaturated fatty acids, specifically eicosapentaenoic (EPA), docosapenaenoic (DHA) and total n-3 fatty acids.

In seeking to rationalize these findings with the differing results outlined above, it should be noted that the organic diet used by Castellini *et al.* (2002) did not consist completely of organic ingredients. The diet used contained 15 percent regular soybean meal, which was permitted temporarily in organic diets at the time the experiment was conducted due to a scarcity of organic feedstuffs. By 2012 the temporary provision in Europe allowing some non-organic feedstuffs to be used in organic production will have been phased out (unless the provision is extended). Consequently the research findings available to date from Europe for poultry (and pigs) may differ from the results obtained after 2012, because it is more difficult to formulate optimum diets for poultry (and pigs) using organic feedstuffs.

Consumer findings

Some consumers purchase whole chickens for meat, others prefer chicken parts. A yellow-skinned or colored-skin bird is preferred by some consumers, others preferring white-skinned birds. These features are taken into account by suppliers in servicing the markets in different regions.

The purchase of organic poultry meat appears to be governed by the same factors governing the purchase of beef and pork. The relative importance of these factors appears to vary according to region or country as exemplified in Scandinavian findings. A study in Denmark showed that 22, 11, 33, 24 and 19 percent, respectively, of consumers purchased organically produced bread, meat, eggs, vegetables and dairy products (Borch, 1999). The corresponding figures were 13, 12, 19, 19 and 13 percent of consumers in Sweden and 11, 9, 17, 16 and 11 percent in Norway. The number of consumers in the three countries who never purchased organic food was 33, 35 and 49 percent, respectively. There were considerable differences among the three countries in the stated reasons for organic purchases, with Danish and Norwegian consumers stating that the main reason for buying organic food was their belief that it was healthier and of better quality than intensively produced food, whereas the main motive for Swedish consumers was their concern for the environment and animal welfare. Swedish and Norwegian consumers trusted the accuracy of the organic trademark, although the Danes were more skeptical. Consumers in all three countries were willing to pay a higher price for organically produced food.

Consumers in the United States also are increasingly interested in both free-range and organic chicken (Alvarado *et al.*, 2005). In a study conducted by these authors, consumers were asked to compare the eating quality and shelf-life of meat from organic free-range broilers with meat from conventionally raised broilers. Free-range chicken breasts were significantly larger (153 g) than conventional chicken breasts (121 g), and this was attributed to increased exercise and increased age of the bird. No significant differences in breast fillet tenderness or composition were noted between the two types of meat. Breast fillets from the free-range birds had higher pH values (5.96 vs. 5.72, respectively) and were darker (49.14 vs. 53.46 units, respectively) than fillets from the conventionally raised birds. Free-range fillets had significantly higher APC (aerobic plate count) and coliform count and exhibited signs of spoilage earlier than conventional fillets, findings of potential consumer importance. Consumers found no difference in fillet juiciness, tenderness or flavor. Conventional breast fillets, however, were preferred to free-range fillets. Trained panelists found no difference in tenderness or flavor of drumsticks, but found meat from free-range birds to be juicier and with a stronger attachment of the meat to the bone. Conventional and free-range chicken had many similarities in meat quality and sensory attributes, but meat from free-range poultry had a shorter shelf-life than meat from commercially raised chickens, confirming other findings related to the higher content of polyunsaturated fatty acids in organic poultry meat.

Other researchers have reported a shorter shelf-life in organic broiler meat. Lawlor *et al.* (2003) studied changes in chicken burger meat manufactured, cooked and stored in a modified atmosphere at 4°C under fluorescent light for 1–7 days. MDA–TBA values used to monitor oxidative changes were in the following order throughout each storage period: organic>free-range>conventional. It was concluded that cooked breast burgers from broilers fed organic diets had a lower shelf-life (oxidative stability) compared with cooked breast burgers from broilers raised on free-range or conventional systems.

As with other classes of farm stock, the trend in organic chicken production is to use heritage breeds rather than fast-growing hybrids and to grow them to higher weights. The research and consumer findings suggest that the result is a slightly tougher meat but with an enhanced flavor that is preferred by some consumers (Blair, 2008).

Fish

As stated earlier in this book there is a standard for organic fish, at least in Europe. This standard (Regulation 710/2009) is based on farmed fish. It is likely that North America will follow suit and develop comparable regulations: discussions and consultations are currently underway. The main problem in developing a standard in North America is that many in the organic industry view as anathema the designation of farmed fish as "organic" and that wild fish cannot receive the "organic" label.

However, the European decision is logical. No international organization accepts the designation of wild fin fish as organic. This is based on the fact that the producer (harvester) cannot guarantee that the conditions under which the fish were raised were in accordance with organic regulations. The most recent draft update of the basic standards for the International Federation of Organic Agriculture Movements (IFOAM) states that "organisms which are moving freely in open waters, and/or which are not inspectable according to general procedures for organic production, are not covered by these standards." Also, the Codex Alimentarius standards which are developed jointly by the World Health Organization and the Food and Agriculture Organization of the United Nations do not allow fished or hunted species to be certified as organic.

The new rules (2010) in Europe cover organic aquaculture production of fish, shellfish and seaweed. The rules set EU-wide conditions for the aquatic production environment and for the separation of organic and non-organic units and specify animal welfare conditions including maximum stocking densities, a measurable indicator for welfare. The rules specify that biodiversity should be respected, and do not allow the use of induced spawning by artificial hormones. Organic feeds should be used, supplemented by fish feeds derived from sustainably managed fisheries. Commenting on the new rules, the EU Fisheries and Maritime Affairs Commissioner, Maria Damanaki, is quoted as stating:

> Europe-wide rules for organic aquaculture have become a reality. They will give consumers a better choice and are a boost for sound and environmentally acceptable production and a viable alternative to the more traditional intensive approach. The EU is the biggest market in the world for seafood and it is fitting that Europe should play a leading role in establishing comprehensive rules in this domain. Among the priorities for my term are sustainability and social cohesion for the fishing and aquaculture sectors. These new rules for organic aquaculture are a milestone by integrating these priorities into aquaculture.

Research findings

A main difference that has been reported between farm-raised and wild fish is the lipid component. For instance, Alasalvar *et al.* (2002) reported that cultured sea bass contained significantly more fat than wild counterparts. The lipids of cultured sea bass contained significantly higher proportions of 14:0, 20:0, 18:1n-9, 20:1n-9, 22:1n-9, 18:2n-6 and 20:3n-6 fatty acids, and lower proportions of 16:0, 18:0, 20:4n-6, 20:5n-3, 22:4n-3, 22:5n-3 and 22:6n-3 fatty acid residues than wild sea bass. The percentages of total saturated and polyenoic fatty acids as well as the n-3/n-6 ratio were higher in the wild than in cultured sea bass, whereas the corresponding total monoenoic content was lower. No significant differences were noted in the total content of minerals, iron and zinc being predominant elements among 14 minerals analyzed. They constituted 78.2 and 81.6 percent of the total mineral contents in the flesh of cultured and wild sea bass, respectively. Significant differences were also found in aluminum, titanium and vanadium contents. The differences found in body composition were attributed to the different diets available to the fish.

Similarly, Hamilton *et al.* (2005) analyzed farmed Atlantic salmon, wild Pacific salmon, and farmed Atlantic salmon fillets purchased in supermarkets in 16 cities in North America and Europe, and reported that farmed salmon had greater levels of total lipid (average 16.6 percent) than wild salmon (average 6.4 percent). The n-3 to n-6 ratio was about 10 in wild salmon and 3–4 in farmed salmon. The supermarket samples were similar to the farmed salmon from the same region. One of the concerns about the higher lipid content is that contaminant levels could also be higher, since these are associated with lipids. This was confirmed in the study by Hamilton *et al.* (2005), the lipid-adjusted contaminant levels (ng/g lipid) being significantly higher in farmed (F) Atlantic salmon than in wild (P) Pacific salmon ($F = 7.27$, $P = 0.0089$ for toxaphene; $F = 15.39$, $P = 0.0002$ for dioxin; $F > \text{or} = 21.31$, $P < 0.0001$ for dieldrin and PCBs). The higher fat content in the farmed fish was attributed to a high level of total lipid (30–40 percent) in the fish oil/fish meal that is fed to farmed salmon. The relationship between lipid content of the fish and the contaminant level is of relevance to consumers since agencies such as the American Heart Association and the UK Food Standards Agency recommend the weekly consumption of oily fish species for cardio-protective benefits.

As noted later in this book, it is known that Inuit populations have very low rates of heart disease even though their diet is composed mainly of lipid-rich fish and marine mammals. Scientific studies have demonstrated that consuming two servings of fish a week has protective effects on cardiovascular health. The dietary component responsible for the benefit is now known to be omega-3 fatty acids. Within this group eicosapentaenoic (EPA) and docosahexaenoic (DHA) fatty acids are long-chain omega-3 fatty acids that protect against cardiovascular and inflammatory diseases and are essential to brain development.

The origin of the fish and the way in which analyses are carried out have been shown also to affect the lipid and contaminant contents of farmed fish. For instance, Ikonomou *et al.* (2007) compared skinned fillets from known sources of farmed Atlantic, coho and chinook salmon and wild coho, chinook, chum, sockeye and pink salmon. The skin was removed since it contains a substantial content of fat but is not eaten. Analyzing the flesh without skin therefore provides a more accurate analysis of the fish eaten. These researchers found that Atlantic salmon contained higher PCB concentrations (28–38 ng/g) than farmed coho or chinook salmon, and that levels in these latter species were similar to those in wild counterparts (2.8–13.7 ng/g). The PCB levels in Atlantic salmon flesh were, nevertheless, 53- to 71-fold less than the level of concern for human consumption of fish, i.e. 2000 ng/g as established by Health Canada and the US Food and Drug Administration. Similarly, levels of THg and MeHg in all samples were well below the Health Canada guideline (0.5 µg/g) and the US-FDA action level (1.0 µg/g). On average, THg in farmed salmon (0.021 µg/g) was similar to or lower than wild salmon (0.013–0.077 µg/g). Atlantic salmon were a richer source (2.34 g/100 g fillet) of n-3 HUFAs than the other farmed and wild sources of salmon examined (means, 0.39–1.17 g/100 g).

The quality and safety of farmed salmonids (Atlantic salmon and rainbow trout) were the subject of a study conducted in Canada in 2003 and 2004 (Morin, 2006). Salmonids are carnivores and in fish-farming operations are fed a diet of dried feed consisting mainly of fish meal and fish oils. The fish from which the feed is made are generally classified as fatty or

Table 6.12   Total lipid and fatty acid composition of harvested wild and farmed salmon and rainbow trout (Morin, 2006)

	Rainbow trout		Atlantic salmon	
	Farmed	**Wild**	**Farmed**	**Wild**
		Percent		
Total lipids	5.576	0.953	7.421	6.967
Total fatty acids	3.188	0.593	4.022	3.974
PUFA	1.222	0.343	1.635	1.053
MUFA	1.106	0.106	1.474	2.161
SFA	0.861	0.144	1.038	0.760
EPA + DHA	0.731	0.232	0.855	0.749
n-3 HUFA	0.861	0.255	1.066	0.911
n-3 PUFA	0.931	0.268	1.192	0.961
n-6 PUFA	0.291	0.075	0.442	0.092

PUFA: Polyunsaturated fatty acids; MUFA: monounsaturated fatty acids; SFA: saturated fatty acids; EPA: eicosapentaenoic acid; DHA: docosahexaenoic acid; HUFA: highly unsaturated fatty acids.

moderately fatty, depending on the species, and their lipid content varies from 2 to 15 percent. Farmed salmonids have a higher fat content than wild salmonids and thus contain higher levels of omega-3 fatty acids per serving. The diet of wild fish is lower in fat and more limited in quantity than that of farmed fish, which are fed a diet of lipid-rich feed to satiety. Wild fish are also more active, travel greater distances and thus expend more energy, whereas farmed salmonids are confined in a restricted space and expend less energy. In 2003, samples of farmed salmon and rainbow trout were collected in supermarkets. Samples of wild salmon and rainbow trout were obtained from fishermen. In total, 46 farmed salmon and 37 farmed trout and 10 wild salmon and 10 wild trout were analyzed for their fatty acid content and environmental contaminant content.

The results of this study revealed that the fat (total lipids) composition of the flesh of farmed rainbow trout (5576 mg/100 g) is 5.6 times higher than in wild rainbow trout (953 mg/100 g) (Table 6.12). A surprising result was obtained for Atlantic salmon, which had a more or less equivalent fat level for both the farmed (7421 mg/100 g) and wild fish (6967 mg/100 g). Fatty acids made up slightly more than half of the total lipids for all the fish analyzed.

The fatty acid compositions of farmed rainbow trout and farmed Atlantic salmon were found to be very similar, with saturated fatty acids accounting for 26–27 percent, MUFA 33 percent and PUFA 41 percent. The similarity in both species was attributed to a diet of similar composition. On the other hand, there was a more pronounced difference between the wild and farmed congeners of the same species. Wild rainbow trout had the highest proportion of polyunsaturated fatty acids (58.6 percent), while wild

Atlantic salmon had the highest proportion of monounsaturated fatty acids (53.7 percent). The proportion of saturated fatty acids was virtually identical (24–27 percent) in wild and farmed rainbow trout and farmed Atlantic salmon, but it was lower in wild salmon, at 19 percent. Compared with farmed rainbow trout, wild rainbow trout had a higher proportion of polyunsaturated n-3 fatty acids (PUFA), highly unsaturated fatty acids (HUFA) and eicosapentaenoic (EPA) + docosahexaenoic (DHA) fatty acids. This difference was less pronounced between wild salmon and farmed salmon, and salmon contained a lower proportion of n-3 fatty acids than wild trout. The proportion of n-6 PUFA was higher in wild trout (12.5 percent) than in farmed trout (8.5 percent) and, conversely, farmed salmon had a higher proportion of n-3 PUFA (9.8 percent) than wild salmon (2.3 percent).

In general, concentrations of contaminants were very low in both species, whether farmed or wild (Table 6.13). The Canadian guidelines stipulate a maximum concentration of 2 mg/kg of PCBs in fish flesh. The average and maximum PCB concentrations measured in each group of samples did not exceed 0.014 mg/kg and 0.039 mg/kg respectively, which constitutes a very small fraction (less than 2 percent) of the acceptable standard. The average and maximum concentrations of dioxins and furans measured did not exceed 0.150 ng/kg and 0.480 ng/kg respectively, which represents 1 to 3 percent of the maximum standard (15 ng/kg). Mercury was measured at average concentrations, which did not exceed 0.056 mg/kg, and maximum concentrations of 0.090 mg/kg in the four groups of fish sampled, which represents less than 20 percent of the maximum standard of 0.5 mg/kg.

No trace of pesticides (180 different compounds) was detected in the 68 samples of farmed fish analyzed. Two farmed fish yielded a positive result for lead, namely one trout at 0.15 mg/kg and one salmon at 0.19 mg/kg. The Codex and EEC maximum residue limit is 0.2 mg/kg for this metal.

Table 6.13 Average and maximum concentrations of PCBs, dioxins and furans, and mercury in harvested farmed and wild rainbow trout and Atlantic salmon (Morin, 2006)

Contaminants (concentration)	Rainbow trout		Atlantic salmon	
	Farmed	Wild	Farmed	Wild
PCBs (mg/kg = ppm)				
Average	0.006	0.006	0.014	0.006
Maximum	0.013	0.011	0.039	0.017
Dioxins + furans (ng TEQ/kg = ppt)				
Average	0.041	0.098	0.082	0.150
Maximum	0.175	0.285	0.480	0.440
Mercury (mg/kg = ppm)				
Average	0.021	0.045	0.018	0.056
Maximum	0.040	0.090	0.030	0.080

A few samples contained traces of cadmium ranging from 0.02 mg/kg to 0.04 mg/kg.

Since fish are a source of exposure to environmental contaminants, toxicological reference values (TRVs) have been established by various agencies, including Health Canada, in order to prevent the risks associated with contaminants. The results indicate that, with a consumption of seven meals of farmed trout or salmon a week, the TRV is never reached for any of the contaminants. The maximum values calculated are 38.5 percent of the TRV at 0.13 mg/kg for PCBs and 25% of the TRV for dioxins and furans. All the other values are below 15 percent of the TRV.

It is known that the concentration of contaminants in fish depends on the origin of the fish and the fish species, but it is also related mainly to the composition of the diet. The basis for the acceptance of organic status for farmed fish is that, unlike fishermen, fish farmers can control for the presence of toxic contaminants and pathogens in their fish throughout the production process. The diet of wild fish is totally beyond human control. Additionally, fish can be directly exposed to contaminants derived from industrial, agricultural and municipal dumping activities and even indirectly exposed to these through contamination from the sediments of these different activities, particularly in open coastal or marine production systems. Wild sources of aquatic foods have been reported to carry a greater risk of trace metal bioaccumulation than farmed fish sources due to unmanageable, polluted surface waters or sediments and the concentration of these metals in the food chain (Jensen and Greenlees, 1997).

Studies have been conducted on the muscle quality of farmed fish. For instance Glover et al. (2009) compared salmon of farmed, wild and hybrid origin in a simulated aquaculture production cycle. At slaughter, the farmed salmon were over twice the size of wild salmon, whilst hybrids were intermediate. Condition factor was considerably higher in farmed than in wild salmon, with hybrids displaying intermediate or similar values. For most of the quality traits recorded, i.e. fat content, fatty acid composition, skin coloration, flesh texture, blood and muscle pH, only moderate differences were observed among the three groups. This was especially evident when the values were adjusted to a similar fish weight. Nevertheless, observed differences in astaxanthin concentration among the three groups were significant even when the relationship to size was accounted for. It was concluded that the observed differences were genetic in origin, and are either the direct result of selection, or linked with selected traits.

The differing concentration of lipid and lipid components found in farmed fish has been investigated as a possible tool in identifying farmed from wild fish. Axelson et al. (2009) used a data set of 131 salmon samples from several geographical origins and the distributions of 12 fatty acids to construct a Bayesian belief network. Ultimately only the three most important fatty acids (16:1n-7, 18:2n-6, and 22:5n-3) were used. The training data set yielded a prediction error of 0 percent (68/68 farmed; 20/20 wild correct) while the validation data set prediction error was 4.65 percent (32/32 farmed; 9/11 wild correct). Different randomly chosen validation sets yielded similar prediction accuracies. This model was then applied to 30 market (store-bought) samples, 25 of them labeled as farmed and five as wild. The five samples labeled wild were predicted to be wild, while two of the 25 labeled as farmed were actually predicted to be wild by the model.

Consumer findings

Of all the foods available, it appears that consumers have the strongest reactions to farmed fish as opposed to wild fish. As a consequence, Target Corp., America's second-largest discounter after Wal-Mart Stores Inc., in 2010 pulled all farmed salmon from its stores. The retailer stated that it would no longer carry farmed salmon in its fresh, frozen or smoked seafood sections. Other companies have increasingly shifted away from farmed salmon as a result of pressure by consumers and environmentalists, who prefer wild-caught fish. Salmon farms are viewed by these people as hazardous owing to their effect on the environment and the potential dangers of farmed fish escaping and intruding on native salmon.

It is obvious the perception plays an important role in the selection of wild vs. farmed fish by consumers. For instance, Belgian consumers expressed some belief that wild fish have a better taste than farmed fish (Verbeke *et al.*, 2005). This belief was stronger among older consumers and typically resulted from expected or perceived differences with respect to freshness and texture. On the basis of the total survey sample, around 20 percent of the respondents agreed with the statement that wild fish are more nutritious than farmed fish. As a result, the scientific grounds for substantiating consumers' perception of wild fish being more nutritious than farmed fish are practically non-existent. Approximately 28 percent of the respondents in the total survey agreed with the statement that wild fish are "more healthier" [sic] than farmed fish. However, on the item safety, more respondents disagreed (22 percent) than agreed (18 percent) with the statement "Wild fish are safer than farmed fish". Clearly, when discussing fish, health extends beyond safety issues.

It will be interesting to observe the reactions of consumers once (if) they have the choice of "organic" (farmed) or "wild" in stores. In this case the farmed fish will have been grown according to organic principles, including being raised in environments selected to minimize potential pollutants in the surrounding area (something that is impossible to do with wild fish). Some of the organic fish may have been raised in artificial seawater, making control of environmental pollutants highly effective.

Conclusions

Concerns about "mad-cow disease" and about possible hormone, chemical and antibiotic dangers in regular meats are not justified by the known facts. Food regulations in developed countries are now effective in addressing these issues. They also apply to meat from cloned or gene-modified animals. Incorrect handling of meat, especially poultry meat, that leads to contamination with food-poisoning organisms is a recognized threat to the consumer and is of greater importance than whether the meat is from animals raised conventionally or organically.

Monitoring studies conducted by official agencies indicate that the meat supply in North America is not heavily contaminated by chemical and pesticide residues or dioxins and similar environmental contaminants, and that the efforts being made to reduce the levels appear to be succeeding. The levels found in various surveys can be regarded as trace amounts. Similar findings have been reported in other countries. The limited findings available on organic meat indicate that it is highly unlikely that there is any difference in occurrence of chemical residues of drugs, pesticides, antibiotics and/or growth promotants in conventional and organic beef. Consumers not convinced by the available evidence should purchase organic meat, though it is clear that organic meat is not completely devoid of residues.

The United States, Canada and Australia are among countries that permit the use of hormone implants in cattle. Monitoring studies have proved their safety, deeming them safe and presenting no harm to the consumer of meat from treated animals. However, the European Union (EU) does not agree and currently bans the importation of meat from animals that have received hormone implants, a decision that may be based on trading considerations rather than on the basis of scientific evidence. Hormones are not used in pig or poultry production, therefore the question of elevated hormone residues in these meats does not arise, and organic and conventional meats from these animals are similar in content of natural hormones.

Beef animals raised organically grow more slowly and produce leaner carcasses. As a result the meat tends to have less marbling and is less tender. The profile of the fat is altered with organic production (or with grass feeding), with a higher content of PUFAs (in particular CLA) and is regarded as more favorable in terms of human nutrition. Similar findings have been reported with pigs and poultry, the research and consumer findings suggesting that the result is a slightly tougher meat but that it has an enhanced flavor that is preferred by some consumers (probably an age effect since the organic animals take longer to reach market weight). The main difference between organic (farm-raised) and wild fish is a higher content of fat in the organic fish.

Perception about the quality of organic meat plays a large role in its purchase by consumers.

References

Alasalvar, K.D.A., Taylor, E., Zubcov, F., *et al.* (2002). Differentiation of cultured and wild sea bass (*Dicentrarchus labrax*): total lipid content, fatty acid and trace mineral composition. *Food Chemistry* **79**, 145–150.

Alvarado, C.Z., Wenger, E. and O'Keefe, S.F. (2005). Consumer perception of meat quality and shelf-life in commercially raised broilers compared to organic free range broilers. *Poultry Science* **84**, 129.

Anders, S. and Moeser, A. (2008). Assessing the demand for value-based organic meats in Canada: a combined retail and household scanner-data approach. *International Journal of Consumer Studies* **32**, 457–469.

Andersen, H.J., Oksbjerg, N. and Therkildsen, M. (2005). Potential quality control tools in the production of fresh pork, beef and lamb demanded by the European society. *Livestock Production Science* **94**, 105–124.

Axelson, D.E., Standal, I.B., Martinez, I. and Aursand, M. (2009). Classification of wild and farmed salmon using Bayesian belief networks and gas chromatography-derived fatty acid distributions. *Journal of Agricultural and Food Chemistry* **57**, 7634–7639.

Blair, R. (2007). *Nutrition and Feeding of Organic Pigs*. Commonwealth Agricultural Bureaux International (CABI), Wallingford, Oxford.

Blair, R. (2008). *Nutrition and Feeding of Organic Poultry*. CABI, Wallingford, Oxford.

Blanco-Penedo, I., López-Alonso, M., Miranda, M., *et al.* (2010). Non-essential and essential trace element concentrations in meat from cattle reared under organic, intensive or conventional production systems. *Food Additives and Contaminants A* **27**, 36–42.

Borch, L.W. (1999). Consumer groups of organic products in Scandinavia. *Maelkeritidende* **112**, 276–279.

Bredahl, L. and Poulsen, C.S. (2002). Perceptions of pork and modern pig breeding among Danish consumers. Report No. 01/02. Aarhus School of Business, Aarhus, Denmark.

Brown, S.N., Nute, G.R., Baker, A., *et al.* (2008): Aspects of meat and eating quality of broiler chickens reared under standard, maize-fed, free-range or organic systems. *British Poultry Science* **49**, 118–124.

Callaway, T.R., Elder, R.O., Keen, J.E., *et al.* (2003). Forage feeding to reduce preharvest *Escherichia coli* populations in cattle, a review. *Journal of Dairy Science* **86**, 852–860.

Cameron, N.D. and Enser, M.B. (1991). Fatty acid composition of lipid in longissimus dorsi muscle of Duroc and British Landrace pigs and its relationship with eating quality. *Meat Science* **29**, 295–307.

CAST (2006). Safety of meat, milk, and eggs from animals fed crops derived from modern biotechnology. Issue Paper 34. Council for Agricultural Science and Technology, Ames, Iowa.

Castellini, C., Mugnai, C. and Dal Bosco, A. (2002). Effect of organic production system on broiler carcass and meat quality. *Meat Science* **60**, 219–225.

Cava, R., Ruiz, J., Ventanas, J. and Antequera, T. (1999). Oxidative and lipolytic changes during ripening of Iberian hams as affected by feeding regime: extensive feeding and α-tocopheryl acetate supplementation. *Meat Science* **52**, 165–172.

Corcoran, K., Bernués, A., Manrique, E., *et al.* (2001). Current consumer attitudes towards lamb and beef in Europe. *Options Méditerranéennes* **A46** 75–79.

Cui, S.H., Ge, B.L., Zheng, J. and Meng, J.H. (2005). Prevalence and antimicrobial resistance of *Campylobacter* spp. and *Salmonella* serovars in organic chickens from Maryland retail stores. *Applied and Environmental Microbiology* **71**, 4108–4111.

Cummings, T.C. (2011) Personal communication dated February 11, 2011.

DAFF (2010). Australian National Residue Survey 2008–2009. www.daff.gov.au/agriculture-food/nrs/animal2, accessed October 25, 2010.

Davis, M.A. and Conner, D.E. (2000). Incidence of *Campylobacter* from raw, retail poultry products. *Poultry Science* **79**, 54.

Dinh, T.T., Blanton, J.R. Jr, Riley, D.G., *et al.* (2010). Intramuscular fat and fatty acid composition of longissimus muscle from divergent pure breeds of cattle. *Journal of Animal Science* **88**, 756–766.

Doyle, M.E. (2006). *Natural and Organic Foods: Safety Considerations. A Brief Review of the Literature.* FRI Briefings. Food Research Institute, University of Wisconsin–Madison, Madison, WI.

Edwards, S.A. and Casabianca, F. (1997). Perception and reality of product quality from outdoor production systems in Northern and Southern Europe. In: Sorensen, J.T. (ed.) *Livestock Farming Systems – More than Food Production.* EAAP Publication vol. **89** European Association for Animal Production, Wageningen Pers, Wageningen, pp. 145–156.

FDA (2009). Genetically Engineered Animals. US Food and Drug Administration, Washington, DC. http://www.fda.gov/animalveterinary/developmentapprovalprocess/geneticengineering/geneticallyengineeredanimals/default.htm, accessed April 15, 2010.

French, P., Stanton, C., Lawless, F., *et al.* (2000). Fatty acid composition, including cis-9, trans-11 octadecanoic acid, of intramuscular fat from steers offered grazed grass, grass silage or concentrates. *Journal of Animal Science* **78**, 2849–2855.

French, P., O'Riordan, E.G., Monahan, F.J., *et al.* (2001). The eating quality of meat of steers fed grass and/or concentrates. *Meat Science* **57**, 379–386.

Glover, K.A., Otterå, H., Olsen, R.E., *et al.* (2009). A comparison of farmed, wild and hybrid Atlantic salmon (*Salmo salar* L.) reared under farming conditions. *Aquaculture* **286**, 203–210.

Gottlieb, S. and Wheeler, M.B. (2008). Genetically engineered animals and public health. Compelling benefits for health care, nutrition, the environment, and animal welfare. Report, Biotechnology Industry Organization, Washington, DC. http://www.bio.org/foodag/animals/ge_animal_benefits.pdf, accessed March 15, 2010.

Grashorn, M.A. and Serini, C. (2006). Quality of chicken meat from conventional and organic production. Proceedings of the 12th European Poultry Conference, Verona, Italy, 67.

Grunert, K.G. (2006). Future trends and consumer lifestyles with regard to meat consumption. *Meat Science* **74**, 149–160.

Hamilton, M.C., Hites, R.A., Schwager, S.J., *et al.* (2005). Lipid composition and contaminants in farmed and wild salmon. *Environmental Science and Technology* **39**, 8622–8629.

Hansen, L.L., Magnussen, C.C. and Andersen, H.J. (2001). [Meat and eating quality of organically produced pigs] (in Danish). In Økologisk og udendørs svineproduktion. Internal Report. Danish Institute of Agricultural Sciences **145**, 39–40.

Hartmann, S., Lacorn, M. and Steinhart, H. (1998). Natural occurrence of steroid hormones in food. *Food Chemistry* **62**, 7–20.

Havenstein, G.B., Ferket, P.R. and Qureishi, M.A. (2003). Growth, livability, and feed conversion of 1957 versus 2001 broilers when fed representative 1957 and 2001 broiler diets. *Poultry Science* **82**, 1500–1508.

Heaton, K.L., Smith, G.C., Sofos, J.N., *et al.* (1996). Analysis of pork products for chemical residues. *Journal of Muscle Foods* **7**, 213–224.

Heuer, O.E., Pedersen, K., Andersen, J.S. and Madsen, M. (2001). Prevalence and antimicrobial susceptibility of thermophilic *Campylobacter* in organic and conventional broiler flocks. *Letters in Applied Microbiology* **33**, 269–274.

Husak, R.L., Sebranek, J.G. and Bregendahl, K. (2008). A survey of commercially available broilers marketed as organic, free-range, and conventional broilers for cooked meat yields, meat composition, and relative value. *Poultry Science* **87**, 2367–2376.

Ikonomou, M.G., Higgs, D.A., Gibbs, M., *et al.* (2007). Flesh quality of market-size farmed and wild British Columbia salmon. *Environmental Science and Technology* **41**, 437–443.

Ireland National Food Residue Database (2009). Residue Studies Group, Food Safety Department, Ashtown Food Research Centre, Teagasc, Ireland. http://nfrd.teagasc.ie, accessed October 11, 2010.

Jahan, K., Paterson, A. and Piggott, J.R. (2005). Sensory quality in retailed organic, free range and corn-fed chicken breast. *Food Research International* **38**, 495–503.

Jakobsen, K. (1999). Dietary modifications of animal fats: status and future perspectives. *Fett/Lipid* **101**, 475–483.

Jensen, G.L. and Greenlees, K.J. (1997). Public health issues in aquaculture. *Revue Scientifique et Technique de l'Office Iinternational des Epizooties* **16**, 641–651.

Kugler, M. (2006). Mad Cow Disease and Humans. http://rarediseases.about.com/od/rarediseases1/a/vcjd.htm, accessed September 5, 2010.

Lawlor, J.B., Sheehan, E.M., Delahunty, C.M., *et al.* (2003). Oxidative stability of cooked chicken breast burgers obtained from organic, free-range and conventionally reared animals. *International Journal of Poultry Science* **2**, 398–403.

Lindén, A., Andersson, K. and Oskarsson, A. (2001). Cadmium in organic and conventional pig production. *Archives of Environmental Contamination and Toxicology* **40**, 425–431.

McEachern, M.G. and Schröder, M.J.A. (2002). The role of livestock production ethics in consumer values towards meat. *Journal of Agricultural and Environmental Ethics* **15**, 221–237.

Melton, S.L. (1990). Effects of feeds on flavor of red meat: a review. *Journal of Animal Science* **68**, 4421–4436.

Melton, S.L., Amiri, M., Davis, G.W. and Backus, W. R. (1982). Flavor and chemical characteristics of ground beef from grass-, forage-grain- and grain-finished steers. *Journal of Animal Science* **55**, 77–87.

Miotello, S., Bondesan, V., Tagliapietra, F., *et al.* (2009). Meat quality of calves obtained from organic and conventional farming. *Italian Journal of Animal Science* **8**, 213–215.

Morin, R. (2006). Farmed salmonids are an excellent source of omega-3 fatty acids and contain very few residues of environmental contaminants. Fisheries and Oceans Canada Report. www.dfo-mpo.gc.ca/aquaculture/ref/morin_aaq-eng.htm, accessed October 20, 2010.

Muir, P.D., Beaker, J.M. and Bown, M.D. (1998). Effects of forage-and grain-based feeding systems on beef quality: a review. *New Zealand Journal of Agricultural Research* **41**, 623–635.

Nielsen, B.K. and Thamsborg, S.M. (2005). Product quality and livestock systems. Welfare, health and product quality in organic beef production: a Danish perspective. *Livestock Production Science* **94**, 41–50.

Nou, X., Delgado, J., Patel, J.R., *et al.* (2007). Prevalence of *Salmonella*, Campylobacter and Listeria on retail organic and kosher poultry products. *Proceedings Annual Meeting International Association for Food Protection*, 157.

Nowak, B, Von Mueffling, T., Caspari, K. and Hartung, J. (2006). Validation of a method for the detection of virulent *Yersinia enterocolitica* and their distribution in slaughter pigs from conventional and alternative housing systems. *Veterinary Microbiology* **117**, 219–228.

Poulson, C.S., Dhiman, T.R., Ure, A.L., *et al.* (2004). Conjugated linoleic acid content of beef from cattle fed diets containing high grain, CLA, or raised on forages. *Livestock Production Science* **91**, 117–128.

Razminowicz, R.H., Kreuzer, M. and Scheeder, M.R.L. (2006). Quality of retail beef from two grass-based production systems in comparison with conventional beef. *Meat Science* **73**, 351–361.

Reinstein, S., Fox, J.T., Shi, X., *et al.* (2009). Prevalence of *Escherichia coli* O157:H7 in organically and naturally raised beef cattle. *Applied and Environmental Microbiology* **75**, 5421–5423.

Rodenburg, T.B., Van Der Hulst-Van Arkel, M.C. and Kwakkel, R.P. (2004). *Campylobacter* and *Salmonella* infections on organic broiler farms. *NJAS–Wageningen Journal of Life Sciences* **52**, 101–108.

Russo, C. and Preziuso, G. (2005). Carcass and meat quality of organic beef: a brief review. *Animal Breeding Abstracts* **73**, 11N–14N.

Sadek, I.A., Ismail, H.M., Sallam, H.N. and Salem, M. (1998). Survey of hormonal levels in meat and poultry sold in Alexandria, Egypt. *Eastern Mediterranean Health Journal* **4**, 239–243.

Scholderer, J. (2003). The quality of free-range pork: what consumers want. Proceedings MAPP Conference, Middelfart, Denmark, October 7–8, 2003.

Scholderer, J., Nielsen, N.A., Bredahl, L., *et al.* (2004). Organic pork: Consumer quality perceptions. Report No. 02/04. Aarhus School of Business, Aarhus, Denmark.

Smith, G.C., Sofos, J.N., Aaronson, M.J., *et al.* (1994). Incidence of pesticide residues and residues of chemicals specified for testing in US beef by the European Community. *Journal of Muscle Foods* **5**, 271–284.

Smith, G.C., Heaton, K., Sofos, J., *et al.* (1997). Residues of antibiotics, hormones and pesticides in conventional, natural and organic beef. *Journal of Muscle Foods* **8**, 157–172.

Søltoft-Jensen, A.J. (2010). Organic feed results in tough pork chops! Report of Danish Meat Research Institute. Danish Agriculture and Food Council, Copenhagen.

Sundrum, A. (2010). Assessing impacts of organic production on pork and beef quality. *CAB Reviews: Perspectives in Agriculture, Veterinary Science, Nutrition and Natural Resources* **5**, 1–13.

Sundrum, A., Butfering, L., Henning, M. and Hoppenbrock, K.H. (2000). Effects of on-farm diets for organic pig production on performance and carcass quality. *Journal of Animal Science* **78**, 1199–1205.

Symeon, G. K., Kominakis, A., Panopoulou, E. and Rogdakis, E. (2009). Comparison of physicochemical characteristics and consumers' acceptance of breast meat from retailed organic and conventional broiler chickens. *Epitheōrēsē Zōotehnikēs Epistēmēs* **40**, 63–72.

Usborne, W.R. (1994). Natural vs. regular beef. Mimeo Report, Department of Food Science and Human Nutrition, University of Guelph, Guelph, Ontario.

USDA (1996). Food Safety and Inspection Service, Nationwide Broiler Chicken Microbiological Baseline Data Collection Program, July 1994–June 1995. Office of Public Health Science. http://www.fsis.usda.gov/OPHS/baseline/broiler1.pdf, accessed April 21, 2010.

USDA (2009). DIOXIN 08 Survey: Dioxins and Dioxin-Like Compounds in the U.S. Domestic Meat and Poultry Supply. www.fsis.usda.gov/PDF/Dioxin_Report_0605.pdf, accessed April 20, 2010.

USDA (2011). US Department of Agriculture National Residue Program, 2009 Residue Sample Results. http://www.ams.usda.gov/pdp, accessed July 27, 2011.

Van Immerseel, F., Cauwerts, Devriese, K.L.A., *et al.* (2002). Feed additives to control *Salmonella* in poultry. *World's Poultry Science Journal* **58**, 501–513.

Van Overbeke, I, Duchateau, L., De Zutter, L., *et al.* (2006). A comparison survey of organic and conventional broiler chickens for infectious agents affecting health and food safety. *Avian Diseases* **50**, 196–200.

Verbeke, W., Sioen, I., Brunsø, K., *et al.* (2005). Consumer perception versus scientific evidence of farmed and wild fish: exploratory insights from Belgium. *Aquaculture International* **15**, 121–136.

Wood, J.D., Richardson, R.I., Nute, G.R., *et al.* (2003). Effects of fatty acids on meat quality: a review. *Meat Science* **66**, 21–32.

Woodward, B.W. and Fernández, M.I. (1999). Comparison of conventional and organic beef production systems II. Carcass characteristics. *Livestock Production Science* **61**, 225–231.

7 Milk and Milk Products

Milk and dairy products are important food sources in many countries, providing calcium and high quality protein. Therefore it is important that consumers should have confidence in the safety and quality of these products. However, there is concern expressed by some consumers about the quality of milk, related mainly to the possible presence of hormones. Some consumers become upset when they learn that cows may have been injected with a growth hormone to allow them to produce more milk. The fear is that the hormone might be present in the milk and affect human health. This concern may be receding, especially in the current economic climate and with less media attention being paid to it. However, it is an issue that has to be addressed, so that public fears can be either allayed or substantiated.

The hormone issue

In 1993 the US Food and Drug Administration approved a protein called recombinant bovine somatotropin or recombinant bovine growth hormone (rbST) for the stimulation of milk production in dairy cows. The term "recombinant" indicates that it has been developed using gene-splicing techniques, but the gene-spliced and natural versions of the hormone are functionally indistinguishable (FDA, 2009).

The pituitary gland in a cow produces eight hormones, one of which is growth hormone. This group of natural protein hormones is important in controlling milk production as well as other essential metabolic functions. Low levels of the hormone are found in milk from all cows.

Before approval by the Food and Drug Administration, the recombinant form of the hormone (rbST) underwent the longest and most comprehensive regulatory review of any veterinary product in history. The review was chaired by Henry I. Miller, MD, a physician and fellow at Stanford University's Hoover Institution. He had headed the Food and Drug Administration's Office of Biotechnology from 1989 to 1993. Three years before the Administration approved the marketing of milk from supplemented cows, it published a report in the journal *Science* that summarized more than 120

Organic Production and Food Quality, First Edition. Robert Blair.
© 2012 John Wiley & Sons, Ltd. Published 2012 by John Wiley & Sons, Ltd.

studies showing that the milk of cows treated with rbST posed no known risk to human health. These conclusions were affirmed by additional scientific reviews conducted by the US National Institute of Health, the US Congressional Office of Technology Assessment, and the drug regulatory agencies of the United Kingdom, Canada, and the European Union, and by an issues audit conducted by the inspector-general of the US Department of Health and Human Services. These reviews noted that small amounts of bovine somatotropin are found in milk from all cows, supplemented or not. They also pointed out that, like other proteins, bovine somatotropin is digested in the human gut, so that if present in the diet it would be broken down and would no longer possess activity. Even if it is injected into the human bloodstream it has no biological activity. Subsequent studies by academics and government regulatory agencies around the world have confirmed the assessment reached by the US Food and Drug Administration.

An updated review paper on the safety of milk from rbST-supplemented cows was published by Raymond *et al.* in 2009. The authors of the review comprised a group of medical, nutritional and animal-science experts. The report recorded that milk from rbST-supplemented cows has been a part of the US food supply since receiving FDA approval over 15 years ago and its use has not been associated with any scientifically documented detrimental effects on human health.

Although the recombinant form of the hormone is allowed in the US, several countries, including Japan, Australia, New Zealand, and Canada, have not yet approved its use. Lack of approval in Canada was not on the basis of a risk to human health but on the basis of the health of the cows, specifically an increased incidence of mastitis (inflammation of the udder). The Canadian study found "no biologically plausible reason for concern about human safety if rbST were to be approved for sale in Canada". The European Union in 1990 declared the use of the recombinant form of hormone as safe, but in 1993 placed a moratorium on its sale by all member countries in Europe. This was turned into a permanent ban starting from January 1, 2000, based on considerations of animal welfare and health, similar to the Canadian situation.

Vicini *et al.* (2008) carried out a study to compare the composition of whole organic milk, milk labeled "rbST-free" and regular milk purchased at the retail level. Samples (total 334) of all three types of milk were collected from all 48 contiguous states in the USA and tested for bacterial counts, antibiotics, fat, true protein, solids-non-fat and hormone content. The study found minimal differences among the three types of milk (Table 7.1). Conventional milk had a slightly lower bacterial count than organic or rBST-free and lower levels of estradiol and progesterone than organic milk. There were no differences in the level of bovine somatotropin (bST) in the three milks. Approximately 82 percent of the somatotropin values were less than the limit of measurement (0.033 ng/ml) and 72 percent were less than

Table 7.1   Concentrations of hormones (least-squares mean) in retail milk from conventional, rbST-free and organic dairy production systems (Vicini _et al._, 2008)

	Production system[a]			Significance (P value)
	Conventional	**rbST-free**	**Organic**	
Bacterial counts (1 000 cfu/ml)	11	26	22	0.0001
Bovine somatotropin (ng/ml)	0.005[b]	0.042	0.002[b]	0.098
Insulin-like growth factor-1 (ng/ml)	3.12[y]	3.04[y]	2.73[z]	0.001
Progesterone (ng/ml)	12.0[y]	12.8[y]	13.9[z]	0.019
Estradiol (pg/ml)	4.97[y]	6.63[z]	6.40[z]	0.045

[a] Conventionally labeled milk: did not contain any claims about supplementation with recombinant bovine somatotropin (rbST) or organic production practices. rbST-free: Processor claim that cows were not supplemented with rbST. Organic: milk from farms that were certified to meet US Department of Agriculture organic standards.
[b] Least-squares mean is less than assay limit of detection.
[y,z] Values with different superscripts are different ($P < 0.05$) within whole milk.

the limit of detection (0.010 ng/ml) for the assay. Levels of insulin-like growth factor-1 (IGF-1) were similar in conventional milk and rbST-free milk, and a little lower in organic milk.

The available data indicate, therefore, that milk from cows treated with recombinant bovine somatotropin is as safe as milk from cows not treated with this hormone. The organic dairy industry does not, however, use the technology, preferring to produce milk using more traditional methods.

Various dairy companies marketing organic milk or milk from cows not treated with the hormone have attempted to label their products with statements such as "free of hormones", or statements implying that their product was superior. The US Food and Drug Administration has clarified the matter for the consumer, requiring dairies that use an rBST-free label to also print a disclaimer such as "Government studies have shown no significant difference between milk derived from rBST-treated and non rbST-treated cows."

Raw milk

Some opponents of hormone use in dairy cows go one step further and want unprocessed milk "straight from the cow". This is an appealing image but one that can have dangerous consequences. Commercial milk is tested and heat processed before being shipped for sale to the public, for a good reason. Milk may become contaminated with food-poisoning bacteria and may contain the organism that causes tuberculosis in humans. Several cases have occurred recently in North America on this issue, usually ending in fines and other injunctions on the producers involved. Some producers attempt to get round the regulations by selling raw milk labeled for pet use as milk for human consumption. Others have members of the public buy the cow that the milk is from so that they then become the owners and are ostensibly able to do what they want with the milk in question. All of these manipulations of the foods regulations present dangers to the public.

Producers of raw milk may feel that their animals are very healthy and that it is perfectly satisfactory to use the milk "straight from the cow". However, cows need to be inspected very carefully and often, to ensure that they are not suffering from diseases such as tuberculosis that can be spread to humans.

An example of the difficulty in detecting carriers of tuberculosis in dairy herds was demonstrated recently in the UK where it was found that cattle, not badgers, were a main reservoir of bovine tuberculosis. Starting in the 1970s veterinarians there conducted badger culls, arguing that this was necessary to eradicate a reservoir of the disease. However, the evidence for this was weak. Other findings confirmed that cattle, not wildlife, were the reservoir for the disease. These results demonstrate the difficulty that experienced veterinarians have in monitoring and eradicating tuberculosis in cattle. Organic farmers will have even greater difficulty in dealing with the problem and should not put members of the public at risk by offering raw, unprocessed milk.

One factor that may have attracted organic producers to the concept of raw, unprocessed milk is that the husbandry of the cows may be of a higher standard than on conventional dairy farms. A group at the University of Glasgow Veterinary School in Scotland assessed the level of dairy-cow cleanliness and the quality of the milk on organic and conventional farms in the UK (Ellis *et al.*, 2007). They found that cows become dirtier in the transition from summer grazing to winter housing. The type of farming system (organic or conventional) had no effect on cow cleanliness when cows were at grass, but the organic cows were cleaner when the cows were housed during the winter. There was a link between cow cleanliness scores and milk quality, with herds having lower bulk tank somatic cell counts (a measure of microbial contamination) tending to have a better cow cleanliness score. This relationship was strongest for the organic herds. However, there was no significant link between cleanliness score and bactoscan count (another measure of microbial contamination of the milk) or incidence of clinical mastitis in the cows. This study indicates that all farms should keep the cows as clean as possible as part of a mastitis and milk quality control program. It also indicated that organic milk, although possibly containing fewer microbes than regular milk, was not suitable for selling raw to the public and required pasteurization.

There are some reports of organic milk lasting longer than regular milk, leading some consumers to conclude that it is "cleaner" and contains fewer microbes than regular milk. The conclusion is partly correct. Some organic milk is processed using UHT (ultra-high heat treatment) to preserve it and allow it to be shipped over long distances. The UHT process involves heating the milk to 138°C for 2 to 4 seconds, destroying all organisms, therefore is more effective than normal pasteurization. UHT milk does not need to be refrigerated. The UHT process is common in Europe, especially in areas where people have small or no fridges.

Perhaps the last word on this issue should be given to a medical doctor (Gerald Bonham, MD) who wrote in the *Vancouver Sun* (Canada), October 1, 2010, as follows:

> Re: *Raw milk activists try to goad Fraser Health into a food fight*, Sept. 29.
>
> Resistance to pasteurization, chlorination, immunization and food enrichment and fortification are frequent subjects of media stories with a David-and-Goliath theme, pitting believers against professionals backed by government regulation and funding.
>
> This is unfortunate. The first half of the 20th century will go down as the golden age of improved health – a half-century that saw a huge gain in life expectancy and a reduction in disability, mostly due to these public health measures. Control of infectious killers (smallpox, polio, diphtheria, tuberculosis) was achieved and food enrichment and fortification got rid of rickets, scurvy, pellagra and goitre.
>
> Sadly, the freedom to take on personal risks poses risks to others. An outbreak of streptococcus traced to a "certified" raw milk dairy farm near Nanaimo caused non-customers to be infected. Likewise, a raw milk listeriosis outbreak in Calgary infected non-customers. Even a Fraser Valley dairy farmer stopped drinking his own cows' milk after he went through a three-month spell with no memory of any event. One-on-one education, as with the family who had refused polio immunization until one daughter was paralyzed, is not a great way to educate. The media do it much better.
>
> (*Source*: From *Raw milk activists try to goad Fraser Health into a food fight* by Gerald Bonham MD, *Vancouver Sun*, Canada, October 1, 2010. © Gerald Bonham. Reprinted with the kind permission of the author.)

Antibiotic residues

As with other foods produced on commercial farms, a concern is the possible presence of antibiotic residues in milk. US data on this issue are available in the National Milk Drug Residue Database, a program set up in 1991 to compile the results of milk residue testing by industry and regulatory agencies. An independent third party, under contract to FDA, operates the database. The NMDRD is a voluntary industry program, with mandatory reporting by State Regulatory Agencies under the National Conference on Interstate Milk Shipments.

This aspect of food monitoring in the United States is therefore different from that for meat and eggs, which are monitored for safety directly by the USDA through its Food Safety and Inspection Agency.

Fifty states or territories submitted data on samples and tests conducted during fiscal year 2009. The Grade "A" Pasteurized Milk Ordinance (PMO), the rules that State Regulatory Agencies use to implement their Grade "A" milk safety program, requires that all bulk milk tankers be sampled and analyzed for animal drug residues before the milk is processed. Any tanker found positive is rejected for human consumption. During this period 3 311 437 samples from bulk milk tankers were analyzed. Of these samples 861 (0.026 percent) were positive for a residue. The trend since 1996 has been for a steady reduction in the incidence of samples positive for the presence of antibiotics. Farmers are financially liable if antibiotics are found in the milk so they take these regulations very seriously and do not ship the milk from cows that are being treated for illnesses such as udder infections. To date, the database contains information on conventional milk only. No comparable data appear to be available for organic milk. If a cow in an organic herd needs to be treated with antibiotics, she is removed from the dairy herd and not returned to it for a period of up to 12 months.

Since several countries have reported violative levels of residues in imported dairy products, emphasis is now being placed on exporting countries. Occurrence of antibiotic residues in farm and retail milk samples has been studied in Brazil, reflecting a higher incidence of violations in that country. In one study (Nunes and D'Angelino, 2007), 50 samples (20 samples of farm tank milk and 30 samples of retail milk) from the state of Sao Paulo were analyzed. Five of the 50 samples (four farm and one retail sample) tested positive for antibiotic residues. The problem appears to occur in more than one area of the country. In another study Nero *et al.* (2007) tested 210 milk samples of unprocessed milk from four different milk-producing areas in Brazil. Antibiotic residues were detected in 24 samples (11.4 percent), and were found in all regions. The results indicated that antibiotic residues may be an important chemical hazard in milk products in Brazil in contrast to the situation in regions such as North America, Australia, New Zealand or Europe.

Pesticide and chemical residues

Another consumer concern is that milk produced conventionally may contain residues of pesticides and chemicals.

More than half of the 9843 food samples collected by the FDA for monitoring in 1997 were of imported foods. No violative residues were found in domestic milk, dairy products, eggs, bananas, apple juice, grains and grain products, and seafood. Violations were detected in 85 of the 5342 imported foods tested, but no violative residues were found in imported milk, dairy products, eggs, seafood, bananas and apple juice. In 2008, 1398 domestic and 3655 import samples of food were collected and analyzed by the FDA. No violations were found in the milk/dairy products/eggs group for either domestic or import samples.

Testing in Canada is the responsibility of a federal agency, the Canadian Food Inspection Agency. Results of their compliance summary for the 5-year period ending 2004 show a 100 percent compliance rate in domestic milk and dairy products for chemical contamination. Imported samples indicated more of a problem. Recent data from the UK also indicate a 100 percent compliance rate for domestically produced milk. As in North America, the data do not include comparable figures for organic dairy products.

Results from the government food monitoring program in Ireland showed that, over the period 1998–2005, the levels of chemical residues in milk fell. In 2005 a total of 140 samples were tested, with none showing residue levels in violation of the regulations. Mycotoxins (fungal toxins) were found in milk, indicating some contamination of the feed or perhaps bedding.

No comparable reports on organic milk appear to be available; however, the scientific literature contains some reports comparing the levels of chemical residues in regular and organic milk.

The possible presence of contaminants and chemical residues in organic milk was investigated in an Italian study (Ghidini *et al.*, 2005). The study involved 12 (six conventional, six organic) farms, with one milk sample (1000 ml) taken per month from the farm tank. The researchers were careful to couple each organic farm to a conventional one within a range of 2 km in order to cover the same production area. All farms had between 80 and 150 lactating cows. Analyses were conducted for organochlorine pesticides, polychlorinated biphenyls (PCBs), lead, cadmium and mycotoxins in both organic and conventional milk. It was found that the concentrations of pesticide and PCB residues were lower than the legal limits in both organic and conventional milk, and that concentrations of lead and cadmium residues were very low and did not differ between organic and conventional milks (1.85 vs. 1.68; and 0.09 vs. 0.16 µg/liter, respectively). Concentration of the mycotoxin aflatoxin M1 was significantly higher in some samples of organic milk than in conventional milk. A total of 49 percent of the organic samples had concentrations of aflatoxin M1 above the legal limit of 50 ng/liter set by EU Regulation 466/2001. The average value for this mycotoxin in organic and conventional milk was found to be 35 and 21 ng/liter, respectively.

The possibility that synthetic antioxidants added to feedstuffs to prevent rancidity can be transferred to milk was investigated by Pattono *et al.* (2009). In this investigation, samples of conventional ($n = 11$) and organic ($n = 81$) milk, both raw and heat-treated, were analyzed for the presence of synthetic antioxidants (butylated hydroxytoluene, butylated hydroxyanisole, dodecyl gallate, propyl gallate and octyl gallate) to verify whether those labeled as "organic" complied with EU Regulations on the use of additives in such products. The analysis detected only the antioxidant BHT and its aldehyde BHT-CHO in all 11 conventional milk samples and in 18 of 81 organic milk samples. The investigation highlighted the importance of strict control of organic and conventional dairy production, since synthetic antioxidants added to feedstuffs to prevent rancidity can be transferred to milk.

Residues of pesticides were measured in milk (non-organic) in an Italian study (Pagliuca *et al.*, 2006). The aim was to determine the level of contamination of unprocessed milk with the main organophosphate pesticides used in Italy and to develop a risk management plan to deal with any identified problems. Samples were collected at four dairy plants directly from the tanker trucks during delivery, from a total of 920 tonnes of milk. Of 135 samples analyzed, 37 were contaminated at a trace level and 10 showed pesticide contamination levels ranging from 5 to 18 µg/kg. However, no positive sample exceeded the maximum residue level (MRL) set by the European Commission.

A study in Latvia compared the level of chemical pollution in organic and conventional milk (Zagorska *et al.*, 2005). Levels of heavy metals and aflatoxin (a mycotoxin) were measured in a total of nine organic bulk milk and nine conventional bulk milk samples collected from different regions of the country. The average level of lead in organic and conventional samples was 0.024 and 0.031 ppm, respectively, both exceeding the permissible level. The cadmium content in organic and conventional milk samples was very low and similar in both types of milk. The legally accepted upper limits of iron, copper and zinc were not exceeded in any sample. The researchers concluded that there were no significant differences between organic and conventional milk in the levels of aflatoxin and heavy metals.

Levels of chemical elements were measured in milk from conventional and organic farms in Poland (Gabryszuk *et al.*, 2008). One of the conventional farms was based on a grazing system; in the other conventional farm the cows were housed more intensively. Samples were analyzed for 29 chemical elements. Of the potentially toxic chemicals measured, the levels of aluminium and arsenic were similar in all the samples. Cadmium and lead levels were highest in milk from the conventional farm where the cows were allowed to graze, and mercury level was highest in milk from the organic farms. None of the levels found were in violation of the standards.

Nutritional and organoleptic qualities

Research findings

Milk quality is usually defined in terms of solids content, including fat and protein contents, fatty acid composition, protein components, mineral and vitamin contents, somatic cell counts and the effect of these various attributes on the processing quality of the milk. The somatic cell count is a measure of the occurrence of mastitis (udder infection) in the dairy herd, and is used as an index of the keeping quality of the milk.

Several investigations have been carried out to test whether organic milk differs in composition and consumer acceptance from milk produced conventionally. This work is complicated by the fact that breed of cow, stage of lactation, feed composition and a lower production found on organic farms as a result of a lower usage of concentrate supplementation can affect milk composition. Another factor is that feed composition during the year is more likely to fluctuate on organic farms than on conventional farms due to seasonal changes in forage composition. This factor can also influence the composition of the milk. A further factor is that organic milk may be pasteurized by ultra-high heat treatment (UHT) to allow it to be shipped to markets distant from the farm of origin.

Investigations involving raw milk collected at the farm and continued over an entire year are therefore most appropriate in studying changes in milk. One such study was conducted by Toledo *et al.* (2002) in which raw milk samples from 31 organic dairy farms and 19 conventional dairy farms in Sweden were collected once a month for one year. The samples were analyzed for gross composition, somatic cell count, fatty acids, urea, iodine and selenium. The results showed small or no differences (Table 7.2). The only significant differences found were in urea content and somatic cell counts, both of which were lower in organic milk. In addition, levels of selenium (but not iodine) were lower in organic milk, which is of nutritional importance since dairy products are significant dietary sources of selenium in Scandinavia. The authors considered the findings predictable since the gross composition of raw milk is primarily determined by a combination of

Table 7.2 Comparison of average gross composition, urea level and somatic cell count between small/large organic and corresponding conventional dairy farms and average values in milk delivered to Swedish dairies during 1996

	Organic		Conventional		Average values for milk delivered to dairies
	Small herds	Large herds	Small herds	Large herds	
Protein (%)	3.36	3.39	3.38	3.37	3.37
Fat (%)	4.28	4.25	4.32	4.37	4.34
Lactose (%)	4.84	4.87	Not analysed	Not analysed	4.80
Somatic cell count (n/ml)	174	198	205	186	233
Urea (mmol/liter)	3.41	3.96	4.35	4.62	5.00

(*Source*: reproduced from Toledo, P., Andren, A. and Bjorck, L. (2002). Composition of raw milk from sustainable production systems. *International Dairy Journal* 12, 75–80. © 2002 Elsevier Science Ltd. All rights reserved.)

Table 7.3 Average concentrations of nutrients and bacterial counts in retail milk from conventional, rbST-free and organic dairy production systems (Vicini *et al.*, 2008)

	Production system		
	Conventional	rbST-free	Organic
Bacterial counts (1 000 cfu/ml)	11	26	22
Composition			
Fat (g/kg)	33.0	33.8	33.8
Lactose (g/kg)	47.1	47.0	46.7
Protein (g/kg)	31.4	31.5	32.2
Total solids (g/kg)	120.7	121.6	122.0
Solids-non-fat (g/kg)	87.7	87.7	88.2

animal breed and feeding regimen. In Sweden there are no major differences in dairy breeds between conventional or organic dairy farms and the feeding regimen on both conventional and organic farms is characterized by a high proportion of roughage.

Vicini *et al.* (2008) carried out a study in the United States to compare the composition of whole organic milk, milk labeled "rbST-free" and regular milk purchased at the retail level. Samples (total 334) of all three types of milk were collected from all 48 contiguous states and tested for bacterial counts, antibiotics, fat, true protein, and solids-non-fat. The study found minimal differences among the three types of milk (Table 7.3). Conventional milk had a slightly lower bacterial count than organic or rBST-free, suggesting that the organic milk was not UHT. Organic milk had a 2.3 percent higher protein content than the other two types of milk, a statistically significant effect. The researchers speculated that this effect might be due to the lower production found on organic dairy farms or to different breeds being used. Antibiotics were not detectable in any of the milk samples. As a result of the findings the researchers concluded that conventional, rbST-free and organic milk were similar in composition.

Grasses and legumes contain significant levels of polyunsaturated fatty acids (PUFAs), therefore several investigations have been conducted to determine whether feeding high-forage diets to cows has the potential to alter the fatty acid composition of the milk and enhance its value for human consumption. Milk contains a large number of PUFAs in the n-3 (omega-3) fatty acid group and the conjugated linoleic acid (CLA) isomer *cis*-9, *trans*-11 C18:2. The principal n-3 fatty acid in milk is α-linolenic acid (C18:3).

The research findings relating to this issue are not consistent.

Ellis *et al.* (2006) reported on a comparison of the fatty acid composition of organic and conventional milk based on samples taken from bulk collection tanks in the UK. The investigation lasted 12 months and involved 17 organic and 19 conventional dairy farms. All milk samples were analyzed for fatty

Table 7.4   Fatty acid composition (percent of total fatty acids) of conventional and organic milks

FA group type	Milk type	
	Conventional	Organic
Saturated FA	67.25	68.13
Monounsaturated FA	27.63[a]	26.19[b]
Polyunsaturated FA	3.33[a]	3.89[b]
Total n-3 FA	0.66[a]	1.11[b]
Total n-6 FA	1.68	1.68
C18:1 *trans*-11 (vaccenic)	1.75	2.06
C18:2 *cis*-9, *trans*-11 (CLA)	0.58	0.65

[a,b]Different superscripts indicate significant differences between milk types ($P < 0.01$).
(*Source*: Ellis, K.A., Innocent, G., Grove-White, D., *et al.* (2006). Comparing the fatty acid composition of organic and conventional milk. *Journal of Dairy Science* **89**,1938–1950. © 2006.)

acid content, but total fat content was not reported. Included in the fatty acid analyses were saturated fatty acids, the ratio of PUFAs to monounsaturated fatty acids, total n-3 FA, total n-6 FA, conjugated linoleic acid (CLA), and vaccenic acid. The ratio of n-6:n-3 FA was also compared.

The results (Table 7.4) showed that organic milk had a higher proportion of PUFA to monounsaturated fatty acids and of n-3 fatty acids than conventional milk, and had a consistently lower ratio of n-6:n-3 fatty acids than conventional milk. There was no difference between organic and conventional milk with respect to the contents of CLA or vaccenic acid. A number of factors other than farm type were identified as affecting milk fatty acid content, including month of year, herd average milk yield, breed type, use of a total mixed ration, and access to fresh grazing. It was concluded that organic dairy farms in the UK produce milk with a higher average content of PUFAs, particularly n-3 fatty acids, throughout the year.

A recent study of UK retail milk found significant differences in fat composition between organic and conventional milk (Butler *et al.*, 2011). The study involved 22 retail brands of milk, 10 being organic. On average, organic milk was higher in fat than conventional milk (3.75 vs. 3.49 percent) and had a different FA profile, with PUFA 39.4 vs. 31.8 g/kg of total FA, CLA 97.4 vs. 5.6 g/kg of FA, and α-linolenic acid 6.9 vs. 4.4 g/kg of FA. The difference in fat composition was greater for summer than for winter milk, attributed to changes in forage availability, quality, and intake.

Somewhat similar findings on the effect of organic feeding on the fatty acid profile in milk was reported by Tsiplakou *et al.* (2010). This investigation involved conventional and organic dairy sheep and goats. It was found that the fat content was lower in the organic milks but had a different fatty acid profile. Compared with milk from conventionally raised animals the milk from organic sheep had higher contents of MUFA, PUFA, α-linolenic acid (LNA), *cis*-9,*trans*-11 CLA, and ω-3 FA, and the milk from organic goats had higher contents of α-LNA and ω-3 FA.

On the other hand, a recent study at Cornell University, New York, did not find differences of similar magnitude (O'Donnell *et al.*, 2010). The milk was obtained as retail samples from the 48 contiguous states as conventionally produced milk with no specialty labeling, milk labeled rbST-free, or milk labeled organic. All the milk had been pasteurized. No statistically significant differences were found in the fatty acid composition of conventional and rbST-free milks. However, these milks were statistically different from organic milk for several fatty acids. As a percentage of total fatty acids, organic milk was higher in saturated fatty acids (65.9 vs. 62.8 percent) and lower in monounsaturated fatty acids (26.8 vs. 29.7 percent) and polyunsaturated fatty acids (4.3 vs. 4.8 percent) than conventional and rbST-free retail milk samples. Likewise, among bioactive fatty acids, organic milk was slightly lower in *trans* 18:1 (2.8 vs. 3.1 percent) and higher in n-3 (0.82 vs. 0.50 percent) and CLA (0.70 vs. 0.57 percent) when the results were expressed as a percentage of total fatty acids. The researchers pointed out that, from a public health perspective, the trend for some of these differences would be considered desirable and for others would be considered undesirable. However, without exception, the differences in milk fatty acid composition among the milk types studied were considered minor and of no physiological importance in terms of public health or dietary recommendations. Their overall conclusion was that, when the data were combined with previous analytical comparisons of the quality and composition of these retail milk samples, the results indicated that all of these milks were of similar nutritional quality and wholesomeness.

The explanation for the disparity in the findings is probably related to the amount and type of forage used to feed the cows. High-yielding cows need to be fed a higher proportion of concentrate to forage in the ration than low-yielding cows, which affects the fatty acid composition of the milk fat. Both grasses and legumes, which are used as forages in cattle feeding, contain significant levels of PUFAs. However, harvesting these crops as hay or silage rather than feeding them fresh results in a reduction in the CLA content. Investigations cited by Weller and Bowling (2007) showed that changing the feed from conserved forages to fresh herbage increases the CLA concentration in the milk. There are also differences between crops. Dewhurst *et al.* (2003) fed dairy cows either grass or legume silages and reported improved intakes and milk yields with the legume silages (lucerne, red clover, white clover), also higher concentrations of PUFAs in milk, particularly α-linolenic acid. The highest concentrations were found in the milk from cows fed red clover silage. Corn has a higher CLA content than grass and as a result the linoleic acid content of milk from cows fed maize silage is higher than that in milk from cows fed grass silage diets, with the total PUFA concentrations being similar. The CLA content of milk is also influenced by the breed of cow, with milk from Jersey cows having a lower concentration than milk from either Friesian or Holstein cows.

Seasonal variation in fatty acid composition of milk is regarded as being due to forage growth pattern, succulent growth in the spring and fall being associated with high intakes of forage and an increased content of PUFA. These influences, combined with a change in rumen fermentation patterns in high-yielding cows as well as a genetic difference compared with lower yielding cows, are likely to result in changes in the fatty acid profile of the milk.

Hermansen *et al.* (2005) compared the contents of major and trace elements in organically or conventionally produced milk. Concentrations of aluminium, copper, iron, molybdenum, rubidium, selenium and zinc were within published ranges. Concentrations of arsenic, cadmium, chromium, manganese and lead were lower, and concentrations of cobalt and strontium were higher than published ranges. The organic milk had a slightly lower content of calcium (1.16 vs. 1.17 g/kg), phosphorus (1.06 vs. 1.10 g/kg) and magnesium (1.06 vs. 1.10 g/kg) but the differences were not significant statistically The organically produced milk, compared with conventionally produced milk, contained a significantly higher concentration of molybdenum (48 vs. 37 ng/g) and a lower concentration of barium (43 vs. 62 ng/g), europium (4 vs. 7 ng/g), manganese (16 vs. 20 ng/g) and zinc (4400 vs. 5150 ng/g) respectively.

Results to date suggest that the nutritional content of organic milk is similar to that of conventional milk. There may be a different profile of fatty acids in organic milk, with a higher proportion of PUFA relative to other fatty acids, but this effect does not appear to be consistent. This difference will be smaller in fat-reduced milk.

Consumer findings

The published literature on factors influencing consumer choice of organic milk is limited. Part of the reason for this is that milk is a much more uniform product than other products such as beef. Milk has to be marketed to the public largely through milk boards which ensure that the product is heat-treated to ensure its safety from microbial diseases for the human consumer. The process may also remove some of the fat, which, in view of the strong consumer demand for low-fat foods, may also add to its appeal.

Hill and Lynchehaun (2002) found that the main reasons stated for buying organic milk were health and better taste and that it was felt to be better for the environment. The health issue was more pronounced in families with children in the household. Consumers also perceived organic food to be more nutritious than conventional food. Price was the primary reason mentioned by consumers for not purchasing organic milk, as it was felt to be quite expensive. The organic milk in that study was about 25 percent higher in price than standard milk. Mixed opinions were expressed about whether organic milk tasted differently from conventional milk. Some organic consumers bought organic milk because it tasted nicer. Others did not like the taste. Therefore lack of improved taste was identified as the second main reason after price for not buying organic milk.

One factor complicating the taste issue is that, as described above, organic milk is often processed using ultra-heat treatment (UHT) which gives it a slight nutty taste that some consumers like and others dislike. As a result, the results of taste tests are not clear-cut.

Hill and Lynchehaun (2002) concluded that there was a lack of consistency among consumers about the taste of organic milk.

Organic feed is more likely to affect the sensory properties of milk than conventional feed, due particularly to the presence of legumes. For instance, Bertilsson and Murphy (2003) compared the effects of feeding red clover, white clover, lucerne and grass silages to dairy cows and reported that the presence of legumes in the feed, particularly red clover, had a negative effect on the organoleptic quality of milk. According to the authors the taste of this milk "deviated more frequently from what was expressed as 'good quality milk'".

Al-Mabruk *et al.* (2004) found increased oxidative deterioration of milk produced from cows fed red clover silage. Mogensen *et al.* (2010) reported that milk from cows fed a diet containing toasted field beans and a high content of corn had a sour feed odor, a bitter taste and a reduced fatty mouth-feel. In comparison, milk from cows fed a high amount of corn and untreated field beans had a higher sugar-sweet taste and fatty mouth-feel and a lower astringent aftertaste and creamy flavor.

The sensory properties of milk are known also to be greatly influenced by the fat content of the milk, therefore the most meaningful tests are those conducted on whole milk in the raw (unpasteurized) state.

In the USA many of the consumers who purchase organic milk do so to avoid milk from cows treated with BST (Dhar and Foltz, 2005). For these consumers the other attributes are secondary. Consumer preference for organic milk has also been analyzed in other studies. Wang and Sun (2003) analyzed the purchase of organic milk in Vermont and found that price and location were important determinants of purchasing decisions by consumers. Dhar and Foltz (2005) found significant consumer preferences among US consumers for organic milk and, to a lesser extent, rBST-free milk. Results from Japan indicated that public perceptions of safety of organic milk, better taste, an environmentally friendly production process, and health and comfort of the cows are important factors influencing purchasing decisions by consumers. Price was identified as a key inhibitor of consumer demand for organic milk, especially among older consumers (Managi *et al.*, 2008).

Table 7.5   Average consumer assessment[†] of milk from cows fed organically, conventionally or on a pasture-based system

Milk type	Overall liking	Overall flavor	Overall appearance	Overall mouth-feel
Organic	4.67[b]	4.48[b]	5.34[b]	4.92[b]
Pasture-based	5.72[a]	5.71[a]	5.87[a]	5.91[a]
Conventional	5.84[a]	5.94[a]	5.67[b]	5.82[a]

[†]On a scale of 1–9, 1 = dislike extremely to 9 = like extremely. [ab] Means within a column with different superscripts are significantly different at $P < 0.05$.

(*Source*: Valverde, L.P. (2007). Comparison of sensory characteristics, and instrumental flavor compounds analysis of milk produced by three production methods. Unpublished MS thesis, Faculty of the Graduate School, University of Missouri-Columbia, USA. © Laura Valverde. Reprinted with kind permission of the author.)

It is not clear from the results of the Japanese study (or from other studies cited here) whether the organic milk was UHT or not. The UHT process can mask the taste of the initial product. As indicated above, some consumers like the taste of UHT milk and others do not.

One study, in which UHT treatment did not complicate the design or findings, found that the preference for organic milk was lower than for conventional milk (Valverde, 2007). The comparison involved whole (non-defatted) milk from organic, pasture-based and conventional dairy farms. All samples were homogenized commercially, pasteurized and bottled in glass containers except for one sample of organic milk and one sample of conventional milk, which were purchased raw directly from the farmer. Results of a preference test are shown in Table 7.5.

Organic milk was significantly different from conventional milk and milk from pasture-fed cows for the overall liking, overall flavor and overall mouth-feel and also from milk from pasture-fed cows, but not conventional milk, for overall appearance. From the results Valverde (2007) concluded that panelists clearly differentiated organic milk from conventional cows and milk from pasture-fed cows based on their liking. Differentiation between conventional milk and milk from pasture-fed cows was achieved for appearance only. Organic milk was the least liked among the samples, whereas conventional milk and milk from pasture-fed cows were rated similarly.

A triangle test was also used, to determine whether consumers could discriminate between samples of milk from the three dairy production systems. A total of 30 untrained panelists evaluated the milk samples in three consecutive sets of triangle tests. The sample combinations in the triangle tests were: organic milk vs. conventional milk, organic milk vs. milk from pasture-fed cows, and conventional milk vs. milk from pasture-fed cows. Results are shown in Table 7.6.

According to the results shown in Table 7.6, consumers were able to discriminate significantly between organic and conventional milk, and

Table 7.6   Ability of consumers to discriminate between samples of milk produced by cows fed organically, conventionally or on a pasture-based system

Number[†]	Comparison		
	Organic vs. conventional	Organic vs. pasture-based	Pasture-based vs. conventional
Incorrect	10	12	17
Correct	20*	18*	13
Total	30	30	30

[†]Based on using 30 panelists, 17 would have to select correctly the sample that was different to establish statistical significance (P < 0.05). *Significantly different at P < 0.01. (*Source*: Valverde, L.P. (2007). Comparison of sensory characteristics, and instrumental flavor compounds analysis of milk produced by three production methods. Unpublished MS thesis, Faculty of the Graduate School, University of Missouri-Columbia, USA. © Laura Valverde. Reprinted with kind permission of the author.)

between organic milk and milk from pasture-fed cows. They were not able to discriminate between conventional milk and milk from pasture-fed cows. These results were in accordance with the results of the taste test, indicating that organic milk was perceived as milk with lower consumer appeal than milk from pasture-fed cows and conventional milk. Conventional milk and milk from pasture-fed cows received similar scores. The results indicated that there were significant differences among milks from the three different production systems, based on analytical, sensory and discrimination studies. In assessing the value of these findings it should be noted that, although published in a graduate research thesis at the University of Missouri-Columbia, the work has not yet been published in a peer-reviewed journal.

Croissant *et al*. (2007) compared the chemical properties and consumer perception of fluid milk from conventional and pasture-based production systems. Although not a study involving organic milk, it yielded useful information on the effects of a high intake of pasture by cows on the sensory properties of the milk. Fluid milk was collected throughout one growing season from Holstein and Jersey cows located in two herds, one fed a pasture-based diet and one fed a conventional total mixed ration. Milk was batch-pasteurized and homogenized. Instrumental and sensory analyses differentiated the two types of milks, related to a higher concentration of unsaturated fatty acids (including two common isomers of conjugated linoleic acid) in the milk from pasture-based cows. Trained consumer panelists reported a greater intensity of grass and cow/barn flavors in milk from pasture-based cows than in milk from cows fed the total mixed ration. Volatile compound analysis by solid-phase microextraction and gas chromatography–mass spectrometry separated the two types of milk. However, analyses showed no compounds unique to either sample. All identified compounds were common to both samples.

Consumer panelists were unable to consistently differentiate between the two types of milks when evaluated at 7°C, and cow diet had no effect on overall consumer acceptance. These results indicated distinct flavor and compositional differences between milk from cows fed a pasture-based diet and one fed a conventional total mixed ration, but the differences were such that they did not affect consumer acceptance by trained panelists.

Labeling has an important influence on the choice of organic cheese (Napolitano *et al.*, 2010). These researchers compared the expected and actual scores for liking in a test involving organic and conventional Pecorino cheese. The expected scores for liking were found to be significantly higher for organic than for conventional cheese, and for organic cheese the score for expected liking was significantly higher than the actual score, when the test was done under blind conditions. The reverse was found for conventional cheese, the actual score being significantly higher than the expected score. These results indicated that the preference for organic cheese was determined mainly by the information on the label and not by the taste or flavor of the product. The findings are a good example of the "halo effect" produced by an organic label.

Many consumers are willing to pay a higher price for organic products. For example, Millock *et al.* (2002) reported that 59 percent of respondents in Denmark were willing to pay a price premium of 32 percent for organic milk, 41 percent indicated a willingness to pay 40 percent extra for organic potatoes, 51 percent were willing to pay a price premium of 23 percent for organic rye bread, and 41 percent indicated they would pay 19 percent extra for minced organic meat. The findings also indicated that the proportion of respondents willing to pay a price premium decreases as the premium level increases.

Conclusions

The quality of regular milk is extremely well controlled in Western countries and is of high quality, whether from cows treated with bovine somatotropin or not. Reviews of the published findings by medical, nutritional and animal science experts have concluded that milk from rbST-supplemented cows has been a part of the US food supply since receiving FDA approval over 15 years ago and its use has not been associated with any scientifically documented detrimental effects on human health.

The available evidence indicates that regular and organic milk contain similar trace levels of chemical and pesticide residues.

The trend for the last 15–20 years has been for a steady reduction in the incidence of samples positive for the presence of antibiotics. Milk safety programs now require that all bulk milk tankers be sampled and analyzed for animal drug residues before the milk is processed. Any tanker found positive is rejected for human consumption.

Data from several countries indicate a 100 percent compliance rate for domestically produced milk in terms of pesticide and chemical residues. The data from official monitoring agencies do not include comparable figures for organic dairy products.

Some consumers like the taste of organic milk, others do not. The amount of heat used during the processing of organic milk (e.g. UHT) is likely to be a factor influencing the preference for the taste. However, the sale of raw (unpasteurized) to the public is not advised. All milk, whether regular or organic, should be heat-processed before being offered for sale to the public, to avoid the potential spread of diseases.

The higher cost of organic milk is a deterrent to some consumers.

A logical application of these findings for the organic dairy supply industries is to find ways of increasing the amount of organic milk available at a price closer to that of conventional milk, and to determine which components in the organic production system result in milk flavors that are unattractive to some consumers.

References

Al-Mabruk, R.M., Beck, N.F.G. and Dewhurst, R.J. (2004). Effects of silage species and supplemental vitamin E on the oxidative stability of milk. *Journal of Dairy Science* **87**, 406–412.

Bertilsson, J. and Murphy, M. (2003). Effects of feeding clover silages on feed intake, milk production and digestion in dairy cows. *Grass and Forage Science* **58**, 309–322.

Butler, G., Stergiadis, S., Seal, C., *et al.* (2011). Fat composition of organic and conventional retail milk in northeast England. *Journal of Dairy Science* **94**, 24–36.

Croissant, A.E., Washburn, S.P., Dean, L.L. and Drake, M.A. (2007). Chemical properties and consumer perception of fluid milk from conventional and pasture-based production systems. *Journal of Dairy Science* **90**, 4942–4953.

Dewhurst, R.J, Fisher, W.J., Tweed, J.K.S. and Wilkins, R.J. (2003). Comparison of grass and legume silages for milk production. 1. Production responses with different levels of concentrate. *Journal of Dairy Science* **86**, 2598–2611.

Dhar, T. and Foltz, J.D. (2005). Milk by any other name: consumer benefits from labeled milk. *American Journal of Agricultural Economics* **87**, 214–228.

Ellis, K.A., Innocent, G., Grove-White, D., *et al.* (2006). Comparing the fatty acid composition of organic and conventional milk. *Journal of Dairy Science* **89**, 1938–1950.

Ellis, K.A., Innocent, G.T., Mihm, M., *et al.* (2007). Dairy cow cleanliness and milk quality on organic and conventional farms in the UK. *Journal of Dairy Research* **74**, 302–310.

FDA (2009). Report on the US Food and Drug Administration's Review of the Safety of Recombinant Bovine Somatotropin. www.fda.gov/AnimalVeterinary/SafetyHealth/ProductSafetyInformation/ucm130321.htm, accessed July 5, 2010.

Gabryszuk, M., Słoniewski, K. and Sakowski, T. (2008). Macro- and micro-elements in milk and hair of cows from conventional vs. organic farms. *Animal Science Papers and Reports* **26**, 199–209.

Ghidini, S., Zanardi, E., Battaglia, A., *et al.* (2005). Comparison of contaminant and residue levels in organic and conventional milk and meat products from Northern Italy. *Food Additives and Contaminants* **22**, 9–14.

Hermansen, J.E., Badsberg, J.H., Kristensen, T. and Gundersen, V. (2005). Major and trace elements in organically or conventionally produced milk. *Journal of Dairy Research* **72**, 362–368.

Hill, H. and Lynchehaun, F. (2002). Organic milk: attitudes and consumption patterns. *British Food Journal* **104**, 526–542.

Managi, S., Yamamoto, Y., Iwamoto, H. and Masuda, K. (2008). Valuing the influence of underlying attitudes and the demand for organic milk in Japan. *Agricultural Economics* **39**, 339–348.

Millock, K., Hansen, L.G., Wier, M. and Andersen, L.M. (2002). Willingness to pay for organic foods: a comparison between survey data and panel data from Denmark. [online]. Proceedings of a Conference on Consumer Demand for Organic Foods, Denmark. Available at: http://www.akf.dk/organic-foods/conference/willingness.pdf, accessed April 9, 2010.

Mogensen, L., Steensig, J., Vestergaard, J., *et al.* (2010). Effect of toasting field beans and of grass–clover: maize silage ratio on milk production, milk composition and sensory quality of milk. *Livestock Science* **128**, 123–132.

Napolitano, F., Braghieri, A., Piasentier, E., *et al.* (2010). Cheese liking and consumer willingness to pay as affected by information about organic production. *Journal of Dairy Research* **77**, 280–286.

Nunes, M.T. and D'Angelino, J.L. (2007). Occurrence of antibiotics residues in farm and ready-to-drink milk samples. *Higiene Alimentar* **21**, 57–61.

O'Donnell, A.M., Spatny, J.L., Vicini, K.P. and Bauman, D.E. (2010). Survey of the fatty acid composition of retail milk differing in label claims based on production management practices. *Journal of Dairy Science* **93**, 1918–1925.

Pagliuca, G., Serraino, A., Gazzotti, T., *et al.* (2006). Organophosphorus pesticides residues in Italian raw milk. *Journal of Dairy Research* **73**, 340–344.

Pattono, D., Battaglini, L.M., Barberio, A., *et al.* (2009). Presence of synthetic antioxidants in organic and conventional milk. *Food Chemistry* **115**, 285–289.

Raymond, R., Bales, C.W., Bauman, D.E., *et al.* (2009). Recombinant bovine somatotropin (rbST): A safety assessment. Paper presented at the Joint Annual Meeting of the American Dairy Science Association, Canadian Society of Animal Science, and American Society of Animal Science, Montreal, Canada, July 14, 2009.

Toledo, P., Andren, A. and Bjorck, L. (2002). Composition of raw milk from sustainable production systems. *International Dairy Journal* **12**, 75–80.

Tsiplakou, E., Kotrotsios, V., Hadjigeorgiou, I. and Zervas, G. (2010). Differences in sheep and goats milk fatty acid profile between conventional and organic farming systems. *Journal of Dairy Research* **77**, 343–349.

Valverde, L.P. (2007). Comparison of sensory characteristics, and instrumental flavor compounds analysis of milk produced by three production methods. Unpublished MS thesis, Faculty of the Graduate School, University of Missouri-Columbia, USA.

Vicini, J., Etherton, T.D., Kris-Etherton, P., *et al.* (2008). Survey of retail milk composition as affected by label claims regarding farm-management practices. *Journal of the American Dietetic Association* **108**, 1198–1203.

Wang, Q. and Sun, J. (2003). Consumer preference and demand for organic food: evidence from a Vermont survey. Proceedings of the Annual Meeting of the American Agricultural Economics Association, July, 2003, pp. 1–12.

Weller, R.F. and Bowling, P.J. (2007). The importance of nutrient balance, cropping strategy and quality of dairy cow diets in sustainable organic systems. *Journal of the Science of Food and Agriculture* **87**, 2768–2773.

Zagorska, J. and Ciprovicča, I. (2005). The comparison of chemical pollution between organic and conventional milk. In: Research for Rural Development, International Scientific Conference Proceedings, Jelgava, Latvia, May 19–22, 2005, pp. 196–198.

8 Eggs

Eggs, like milk, are an important source of nutrients in our diet, providing protein, vitamins, choline, iron and zinc. Eggs are also good sources of antioxidants known to protect the eyes, namely lutein and zeaxanthin (particularly when hens are fed diets based on yellow corn, also known as maize). This is an important issue for people susceptible to developing macular degeneration and eye cataracts.

However, consumers have several major concerns about eggs, in addition to the usual worries about chemical and pesticide residues relating to all foods. Until recently the main concerns were cholesterol and *Salmonella*. Another concern has now emerged, namely the way in which laying hens are housed, with some consumer resistance to the housing of laying hens in cages. The main benefits of cage-housing are lowered housing costs, ease of operation for the farmer, and reduced contact of the hens with manure, pathogens and predators. However, more and more consumers prefer non-cage systems for egg production.

Free-run or cage-free eggs are produced by hens that are able to move about the floor of the barn and have access to nesting boxes and perches. Free-range eggs are produced in a similar environment to that of cage-free eggs but hens have access to outdoor runs as well. In organic production the birds must be housed in free-run conditions and have open access to outside runs when weather conditions permit. Organic hens cannot be kept in cages.

The downside to allowing hens to run outdoors is that mortality is higher due to increased fighting and there is a higher disease incidence because the birds are more in contact with their manure. The eggs are cleaner when the hens are in cages, an important issue in Europe where eggs are not washed prior to sale. Hens running outdoors are also more liable to pick up toxins in the environment and to become infected with organisms such as *Salmonella*. In addition, the birds are more exposed to diseases such as infectious influenza that can be spread by wild birds and other animals. For that reason all poultry may be required by health authorities to be housed indoors during disease outbreaks, or there may be a requirement that outside runs be screened with a mesh cover.

Organic Production and Food Quality, First Edition. Robert Blair.
© 2012 John Wiley & Sons, Ltd. Published 2012 by John Wiley & Sons, Ltd.

As a result of the controversy about housing, the farming of layers in cages is due to be banned across the European Union by 2012. Cage production is also scheduled to be phased out in California in 2015, following a public ballot in 2009.

What does all of this mean for the consumer? Do concerns about cholesterol, *Salmonella* and residues in eggs justify a switch to organic eggs? The scientific evidence can help consumers make the appropriate choice between regular and organic eggs. Concern about cage or no cage is more of an ethical question and consumers have to make up their own minds on this issue.

The choice between eggs from caged and non-caged hens needs also to take into account the difference in the cost of eggs in the store. Eggs in Europe and in California will cost more as a result of the ban on cages, possibly twice as much. Organic production requirements add further costs, mainly because of the requirement for special feed and a higher mortality rate in the flock.

The situation in California is probably a good example for other regions of how egg production is likely to change in the future following the decision to ban cages. The Californian egg industry produces almost five billion eggs per year from almost 20 million laying hens. The value of production was about $337 million in 2007, down from 1971 when California produced about nine billion eggs annually and was a net exporter. California is now a substantial net importer of eggs from other states, producing about 6 percent of the national total of table eggs and consuming about 12 percent, based on population share. Non-cage production still occurs in California and in other US states. However, the share of non-cage production is quite small, about 5 percent of the total, including the non-cage eggs that also qualify as organic. The effect of the cage ban in California is that cheaper eggs will likely continue to be imported from other states and compete with the more expensive local product.

In those areas where cheaper eggs from caged birds continue to be available, the question for the average consumer is then, "Are the more expensive eggs from non-caged or organic layers worth the extra cost?"

Cholesterol

An elevated blood-cholesterol level is known to be a risk factor for heart disease. In the past eggs were thought to contribute to an increased level of cholesterol in the blood, containing about the same amount of cholesterol as in a single serving of liver, shrimp or duck meat. However, it has now been established that for most people the amount of cholesterol in the diet has only a slight effect on the amount of cholesterol in the blood (Gray and

Griffin, 2009). Most people synthesize a greater amount of cholesterol in their bodies than they take in from their food.

A much greater emphasis is now placed on the type of fat in the diet in relation to high blood cholesterol. As outlined earlier in this book, fat in foods contains different kinds of fatty acids. Some are saturated fatty acids (hard fats), some are unsaturated (fats that are liquid at room temperature and are usually termed oils) and others are hydrogenated (trans) fats which have been converted from liquid to solids by processing. It is now known that the amounts of saturated fat and trans fats in the diet have much greater effects on blood-cholesterol levels than the amount of cholesterol in the food and that they increase the risk of heart disease. For instance, Harvard researchers in a study with more than 80 000 female nurses found that consuming an egg a day did not lead to a higher risk of heart disease (Hu *et al.*, 1999) However, this conclusion appears to apply only to healthy individuals. The nurses in this study who had diabetes and men in another study who had diabetes developed an increased risk of heart disease with higher egg consumption. Another study at the University of Connecticut (Greene *et al.*, 2006) found that 66 percent of elderly men and women showed no significant change in blood cholesterol after eating a diet containing three eggs per day (640 mg cholesterol/day) or an equal amount of a cholesterol-free egg substitute (0 mg cholesterol/day). Thirty-three percent, however, were found to be hyper-responders whose blood cholesterol rose on the egg diet. As a consequence hyper-responders have to be careful about the amount of cholesterol in the diet. Another important finding of that study was that all individuals on the egg diet had an increase in blood lutein and zeaxanthin, a beneficial effect.

Mayo Clinic cardiologist Thomas Behrenbeck (2008) makes the following points about cholesterol and eggs:

> When deciding whether to include eggs in your diet, consider the recommended daily limits on cholesterol in your food ... If you are healthy, it's recommended that you limit your dietary cholesterol intake to less than 300 milligrams (mg) a day ... If you have cardio-vascular disease, diabetes or a high low-density lipoprotein (LDL, or "bad") blood cholesterol level, you should limit your dietary choles-terol intake to less than 200 mg a day.
>
> One large egg has about 213 mg of cholesterol – all of which is found in the yolk. Therefore, if you eat an egg on a given day, it's important to limit other sources of cholesterol for the rest of that day. Consider substituting servings of vegetables for servings of meat, or avoid high-fat dairy products for that day.
>
> If you like eggs but don't want the extra cholesterol, use only the egg whites. Egg whites contain no cholesterol. You may also use cholesterol-free egg substitutes, which are made with egg whites. If you want to reduce cholesterol in a recipe that calls for eggs, use two

egg whites or $\frac{1}{4}$ cup (59 milliliters) cholesterol-free egg substitute in place of one whole egg.

Among the research findings causing the change in opinion about eggs was the Ireland–Boston Heart Study in which researchers followed 600 Irishmen between the ages of 30 and 60 who had lived in Boston for 10 or more years, and their brothers who had never left the old country (Trulson *et al.*, 1990). The Irish brothers ate about twice as many eggs as their American brothers, averaging over 14 per week, yet the Irish brothers had lower blood levels of cholesterol, and their hearts were rated as being from two to six times healthier. The same Harvard doctor examined both groups. An increased physical activity was suggested for the difference in cardiovascular health.

Research published by Harman *et al.* (2008) provided further information about eggs and cholesterol. The study found that people who ate two eggs per day while on a calorie-restricted diet not only lost weight but also reduced their blood cholesterol levels. Two eggs per day were fed to overweight but otherwise healthy volunteers for 12 weeks while they simultaneously followed a reduced-calorie diet prescribed by the British Heart Foundation (BHF). A control group followed the same diet but cut out eggs altogether. Both groups lost between 3 and 4 kg (7–9 lb) in weight and saw a fall in the average level of blood cholesterol. Research leader Dr Bruce Griffin is reported as stating when the findings were publicized: "When blood cholesterol was measured at both six weeks and twelve weeks, both groups showed either no change or a reduction, particularly in their LDL (bad) cholesterol levels, despite the egg group increasing their dietary cholesterol intake to around four times that of the control." Dr Griffin continued: "There is no convincing evidence to link an increased intake of dietary cholesterol or eggs with coronary heart disease through raised blood cholesterol. Indeed, eggs make a nutritional contribution to a healthy, calorie-restricted diet."

Kritchevsky and Kritchevsky (2000) in their review of the epidemiological evidence related to egg consumption and coronary heart disease concluded that no association could be established between egg consumption at levels up to 1 + egg per day and the risk of coronary heart disease in non-diabetic men and women.

This research suggests that many people, including most of those with heart conditions, should be eating more nutritious foods such as eggs. Several researchers have indicated that the high-quality protein found in eggs is beneficial in heart patients when rebuilding damaged heart muscle.

One issue that puzzled nutritionists and health professionals until recently was that Inuit (Eskimos) have low rates of coronary heart disease and cancer despite eating a high-fat diet. For instance, the people of Greenland have a significantly lower rate of mortality from heart attacks than their counterparts in Denmark, yet their diet is traditionally high in fat

(Holub, 2002). One feature of their diet is that it provides up to several grams per day of omega-3 fatty acids in the form of marine mammals (seal, whale), wildfowl (seabirds) and fish. Higher fish intakes in Japan than in North America have also been associated with considerably lower rates of heart disease.

The health-enhancing effect of the Inuit and Japanese diet is now attributed to an increased intake of omega-3 fatty acids, these fats having a positive effect on heart health. Simopoulos (2009) reviewed the recent evidence on omega-3 fatty acids in relation to health and disease in general, not just in relation to heart health. These appear to be important nutrients, which cannot be synthesized in the body.

The result of these findings is that eggs have regained favor with consumers, especially eggs with an altered profile of fats to provide an increased content of omega-3 fatty acids. Hens fed feeds such as flax seed and certain oils have an improved fatty acid composition, with less saturated fat and more polyunsaturated fats (PUFA) such as omega-3 fatty acids, which has a beneficial effect on blood cholesterol level. Organic eggs have these features, hence the increased consumer interest in organic eggs or so-called designer eggs with an increased content of PUFA.

One point to note is that organic eggs do not have a lower content of cholesterol than eggs from hens managed conventionally, e.g. Milinsk *et al.* (2003). Some research indicates they contain more (Minelli *et al.*, 2007).

The conclusion that can be drawn from the above findings is that the dietary contribution of cholesterol from eggs is less important than the amount of saturated and trans fats. Consequently, the choice between regular and organic eggs in terms of cholesterol content is academic and not of major practical importance to the average consumer. The way in which laying hens are housed does not affect the cholesterol level in eggs. However, the feed can be altered to provide a more favorable profile of fatty acids in eggs and this can be done with hens in organic or conventional production.

Salmonella and food-poisoning

A major concern about eggs is the possible presence of *Salmonella* and other food-poisoning germs. *Salmonella* is an organism that can cause serious and sometimes fatal infections in young children, frail or elderly people and others with weakened immune systems. It is the leading cause of human food-borne infections associated with consumption of poultry products world-wide (Van Immerseel *et al.*, 2002). An Italian study showed that the most important concern about eggs in that country was *Salmonella* contamination, particularly in families with young children (Miele and Parisi, 1999). *Campylobacter jejuni* is the most common intestinal bacterial pathogen reported in developed countries and is considered to be of food-borne origin.

The Danish Food Authority announced in 2008 that an EU ruling will soon allow Denmark to ban imports of *Salmonella*-infected food products. By way of background, the Serum Institute in Denmark stated that the worst outbreak of *Salmonella* in 15 years had afflicted thousands of Danes in just a matter of weeks. The Food Authority tested hundreds of food samples to locate the source, likely to be pork or poultry. At the time of the announcement, however, the exact source had not been identified. Denmark was reported to have spent the previous two years documenting the low levels of *Salmonella* in Danish chickens to the Commission, to prove that imported poultry was not as safe as Danish poultry. The statistics presented showed that *Salmonella* was present in one out of every 60 Danish chickens, while one out of every seven imported chickens was infected. In support of the Danish case, Sweden and Finland also have recorded low *Salmonella* levels and were granted permission to refuse *Salmonella*-infected imports when these countries joined the European Union in 1995.

Another response to the *Salmonella* problem was the announcement in 2008 that one of the large egg co-ops in the United Kingdom was now offering a new insurance policy for *Salmonella* and avian influenza. This action was in response to a change in the *Salmonella* rules which could have financial consequences for egg businesses and which came into effect in January, 2009. Where routine testing finds the presence of either *Salmonella enteritidis* or *S. typhimurium*, eggs cannot be used for human consumption unless they have been heat-treated to guarantee the elimination of these strains of *Salmonella* that pose the greatest threat to humans. The financial impact of this restriction of the sale of fresh shell eggs from infected flocks was reported to be a cause for great concern to the co-op, especially for new producers. Therefore this prompted the co-op to introduce the policy, which guaranteed against loss of income from *Salmonella* and avian flu.

These developments demonstrate how seriously the food regulatory authorities take the issue of *Salmonella* contamination in poultry. As a result, the egg industry has changed enormously with improvements in biosecurity and hygiene management on farms. For instance, the UK reported that *Salmonella* contamination rates in UK-produced eggs had fallen by two-thirds since 1996 (Foods Standards Agency, 2004). A survey of 28 518 retail-sold eggs found that just one in every 290 boxes of six eggs on sale had any *Salmonella* contamination, compared with one in 100 in a 1995/1996 survey. It found no difference in contamination rate between free-range and other eggs. All types of retail eggs were included in the survey, with eggs from caged production accounting for 50 per cent of total eggs sampled, free-range eggs 16.9 per cent, barn eggs 16.5 per cent and organic eggs 16.6 per cent. These changes, coupled with the introduction of vaccination of laying hens, have produced a threefold drop in *Salmonella* prevalence in UK-produced eggs. There has also been a decrease in human cases of *Salmonella* food poisoning linked with eggs in the UK.

It is surprising that the vaccine is not yet required in the US, although the UK now claims to have the safest egg supply in Europe as a result of vaccination. A survey conducted by the European Food Safety Agency in 2009 found that about 1 percent of British flocks had *Salmonella* compared with about 60 to 70 percent of flocks elsewhere in Europe.

Vaccination against *Salmonella* infection and other diseases appears to be the obvious answer to medication to combat infections in organic animals. Currently, the rules for organic animal production allow routine vaccination, when "there is an obvious need and other methods of treatment can be regarded as less acceptable" (Berg, 2002). However, some jurisdictions (e.g. Sweden) do not permit the use of vaccines consisting of or manufactured by the use of genetically modified organisms. The EU legislation does allow GMO-based vaccines when there is no other useful alternative available.

Targets for the reduction of *Salmonella* in laying hens, supported by the European Food Safety Authority, have also been recently introduced in the EU and should lead to a lower prevalence of *Salmonella* in eggs. Each member state has to work towards reducing the number of layer-hen holdings contaminated with *Salmonella* by a specific minimum percentage each year, with steeper targets for member states with higher levels of *Salmonella*. The first target deadline was set for 2008. There will also be mandatory vaccination of layer hens against *Salmonella* from 2008 onwards for layer-hen holdings in member states with a *Salmonella* prevalence of 10 percent or more.

There are numerous field reports relating cases of food poisoning in North America to the consumption of uncooked eggs. For instance the International Food Safety Network (Chapman, 2008) reported cases of *Salmonella* infection linked to raw eggs used in mayonnaise, various dishes containing raw eggs and an aioli salad dressing and dipping sauce that was made from raw eggs. Some of the people affected had to be hospitalized. The obvious response to this finding is to cook all eggs.

In 2010 the US Centers of Disease Control and Prevention (CDC) investigated an outbreak of *Salmonella enteriditis* linked to shell eggs. The outbreak began in May and ended in November 2010, affecting almost 2000 consumers in several US states. The investigations in California, Colorado and Minnesota revealed several restaurants or events where more than one ill person with the outbreak strain had eaten. California health officials confirmed that the outbreak had been tracked to in-shell eggs from a single farm in Galt, Iowa. It took until November 30, 2010, for clearance to be given by the FDA for shipments of eggs to resume (CDC, 2010).

It is clear that simply washing the eggs does not remove the risk. The egg may be contaminated internally. Also, washing them might increase the problem. Washing eggs in water colder than the eggs causes the air sac in the egg to contract. As a result the wash-water, including soap and fecal material, can be drawn through the shell and into the egg. Any bacteria present will then have ample opportunity to grow inside the egg and present a potential problem to the consumer if eaten uncooked. Eggs need to be cooked to 145 °F (62.8 °C) for 15 seconds to reduce the risk from *Salmonella*, and pasteurized eggs should be used as a replacement for raw egg dishes.

An important question for the consumer is whether switching to organic eggs reduces the risk of food poisoning from *Salmonella* and related organisms. In attempting to address this issue it is important to review how poultry become infected with these organisms. It is also important to recognize that although the birds may be infected, the organisms may not be transmitted to the egg, or at least not to the interior of the egg.

Poultry can become infected from sources such as bedding, litter, manure, soil, insects, rodent infestations, etc., and the most serious serotypes are those that can pass from the intestine of poultry into the tissues to contaminate the meat and eggs. Prevention of infection by appropriate management protocols, including proper hygiene, is the most important control measure. Feed-related control measures that are known to be effective in helping to control *Salmonella* contamination levels include steam-pelleting of the feed, also the inclusion of approved additives such as prebiotics, probiotics and short-chain fatty acids in the feed mixture.

It is known that eggs from caged birds are less likely to be contaminated with *Salmonella* and related organisms than those from floor-housed birds. This is because the droppings fall though the cage floor and do not come into contact with the eggs. Also, birds that are allowed to range outdoors are more likely to become contaminated from soil, insects and rodent infestation. Therefore, organic eggs are more likely to be contaminated than regular eggs since cage-housing is not permitted in organic production. However, the results of the 2004 UK survey described earlier in this chapter (Foods Standards Agency, 2004) do not indicate more or less of a risk in organic eggs.

Research findings in Finland (Sulonen *et al.*, 2007) reported that 76–84 percent of organic layer farms in Finland were positive for *Campylobacter* contamination, based on examination of manure samples. However, only one of 360 eggs sampled showed contamination of the shell, and no contamination was found in the yolks. In The Netherlands, a study on 31 organic farms (Rodenburg *et al.*, 2004) showed a prevalence of 13 percent for *Salmonella* and 35 percent for *Campylobacter*. The incidence of *Salmonella* was lower and that of *Campylobacter* higher in organic than in conventional broiler flocks.

De Reu *et al.* (2009) compared the bacteriological contamination of eggs from cage and non-cage systems. On average, they found a higher initial eggshell contamination with a total count of aerobic bacteria for eggs from non-cage systems compared with conventional cage systems (5.46 compared with 5.08 log CFU/eggshell, respectively. However, they reported that the major differences found in experimental studies between cage- and non-cage systems are less pronounced under commercial conditions, and that recent research does not indicate large differences in egg content contamination between eggs from cage- and non-cage systems (ignoring outside nest and floor eggs).

Scientists from the National Veterinary Institute in Sweden investigated the causes of death in 914 birds from 172 free-range, barn-reared and battery flocks. They found higher death rates in the free-range and barn-reared flocks than in the battery hen flocks. Most birds died of bacterial infection, commonly *E. coli*. Bacterial infection rates ran at 74 per cent in free-range flocks, 73 per cent in litter-based flocks and 65 per cent in caged flocks (Fossum *et al.*, 2009).

These and other findings show that eggs pose less of a risk to the human consumer from *Campylobacter* than poultry meat.

The US Food and Drug Administration, on March 20, 2009, reported the recall of organic eggs for possible *Salmonella* contamination. The recall involved organic brown eggs produced at a farm in California which distributes eggs to supermarkets in northern California and western Nevada. The recall was initiated after it was determined that the eggs had tested positive for *Salmonella* during an internal investigation by the farm in question, according to the FDA. However, the FDA reported that no known illnesses had occurred in connection with these eggs.

The food monitoring and research findings outlined here indicate that the chances of eggs being contaminated are very low, but that eggs cannot be guaranteed to be free of *Salmonella* or other food-poisoning organisms whatever the source or type. Switching to organic eggs does not remove the problem and may increase it. This conclusion is particularly important for vulnerable groups, such as the elderly, babies and toddlers, pregnant women, and people who are already unwell and more vulnerable to infection. These groups should continue to ensure that eggs from any source are thoroughly cooked before being eaten, otherwise there is a risk of food poisoning. Cooking eggs properly will kill any bacteria present.

Contamination with residues

Antibiotics

In the past, antibiotics were used at low levels in poultry feed to enhance the health and productivity of flocks. The practice was known to result in small residues of some antibiotics in eggs, although most of the antibiotic ingested by the hen was excreted in the manure. When concerns began to be raised about the practice possibly resulting in the emergence of antibiotic-resistant strains of bacteria in humans, the regulations were changed to permit only those antibiotics not used in human medicine to be allowed in feed.

In addition, those approved for use in feed were ones that were not absorbed into the tissues of the bird and acted only in the gut.

The use of antibiotics for this purpose was strictly regulated by the various feed regulatory authorities, e.g. the FDA and the United States Department of Agriculture (USDA) in the US, to ensure their safety and efficacy.

Most countries have now followed the example of Sweden which in 1986 prohibited the use of all antibiotics in feeds. Other countries in the European Union introduced a similar ban as of January 1, 2006. As a result, antibiotics are no longer used on a continuous basis in the egg industry in most countries. Antibiotics that continue to be used in poultry production are permitted under veterinary supervision to treat clinical disease outbreaks. Eggs from treated hens cannot be offered for sale until the completion of a withdrawal period. This is to ensure that no residues are present in the eggs.

The egg industry is adapting to the removal of antibiotics in feeds in other ways, such as improving the hygiene on farms. It has been shown that under ideal production conditions it is possible to achieve good production without the continuous use of antibiotics in feeds. In addition safer alternatives to antibiotics have been introduced, including enzymes, prebiotics, and probiotics.

Results from Australia on monitoring of eggs for residues are typical of those reported from other countries. The most recently available government report from Australia (DAFF, 2010) involved the testing of eggs for a wide spectrum of possible residues of antibiotics. Zero residues were found. Therefore, for the consumer the term "antibiotic-free" in relation to eggs does not mean very much. All commercial poultry – free-range, barn or caged – are fed diets free of antibiotics unless administered by a veterinarian and, if the hens are treated, the eggs are not sent to market.

Consumers still concerned about antibiotic residues in eggs could buy organic eggs. However, this decision has to be viewed as being a personal preference and is not supported by the documented evidence on problems arising from antibiotic contamination.

Chemical residues

As noted above, a trend in many countries is for cage-housing of layers to be phased out, bringing production closer to the organic system which bans cage-housing. This makes academic any possible benefits from cage-housing in terms of reduced contamination of eggs.

There is no doubt that in certain parts of Europe residues in eggs present a risk to the consumer. However, this is not the case in other countries for which documented data are available. Eggs are monitored in several countries as part of the food safety programs. Fewer and less frequent data are available on eggs from these official sources than for other foods, probably because the official monitoring programs place emphasis on foods likely to be more problematic.

Australian data show a reassuring picture of residue data in eggs in that country. Tests on 75 samples and involving 1025 analyses in 2006/2007 showed a 97.3 percent compliance with the Australian standards for chemical and pesticide residues (DAFF, 2010). The results published in 2009 for the 2007/2008 year show a slightly lower compliance rate – 92 percent.

The data collected during 2005 and 2006 in the USDA Pesticide Monitoring Program do not indicate a problem with eggs produced in the USA (USDA, 2010). In those years 100 percent of eggs tested were in compliance with established residue tolerances and formal action levels. However, whereas 100 percent of imported eggs were in compliance with established residue tolerances in 2005, only 57.1 percent were in compliance in 2006. These results are in keeping with the US record on residues in eggs, which has been consistently good for several years now. The survey conducted by the FDA in 1997 found violative residues in domestic foods in 16 of 1171 (1.2 percent) fruits and 41 of 1707 (2.4 percent) vegetables. No violative residues were found in milk, dairy products, eggs, bananas, apple juice, grains and grain products, and seafood. Among the imported food samples, 3 of 322 (0.9 percent) grain and grain products, 24 of 2034 (1.2 percent) fruits, 50 of 2356 (2.1 percent) vegetables, and 7 of 268 (2.6 percent) samples of other products such as nuts and spices had violative residues. No violative residues were found in imported milk, dairy products, eggs, seafood, bananas and apple juice.

The US Food and Drug Administration announced in June, 1999, that all egg and egg-containing products from Belgium, France and The Netherlands were to be detained at ports of entry. The order applied to all imports of eggs, products containing eggs, and game meats (FDA regulated), and all animal products including animal-derived medicated and non- medicated feeds, feed ingredients, and pet foods. These products were detained because of the possibility of contamination with polychlorinated biphenyls (PCBs) and dioxins. The problem was traced to contaminated fat that had been sold to European animal feed manufacturers. The action testified to the effectiveness of the food monitoring system in the US.

Results from Ireland indicate a low level of contamination in eggs in that country (Table 8.1).

Additionally, the monitoring data produced by the Canadian Food Inspection Agency (Table 8.2) present a reassuring picture of residues in eggs (CFIA, 2006). The survey included pesticides, agricultural chemicals, veterinary drugs, environmental pollutants and other impurities, over a 5-year period.

One aspect of the recorded data is that the amount of testing by government agencies is relatively low. Consumers would like to see an increased program of testing.

Table 8.1   Results from the Irish Food Monitoring Program for residues in eggs (2005 results, from Ireland, 2009)

	Hormone-type		Pesticide-type	
	Total samples	Samples non-compliant	Total samples	Samples non-compliant
Eggs	95	0	183	0

(*Source*: adapted from M. Danaher, A. M. Sherry and J. O'Mahony (2010). National Food Residue Database Report 2009. Teagasc Food Research Centre, Ashdown, Dublin, Ireland. Copyright Teagasc 2010.)

These results indicate that, in general, eggs contain zero or merely trace levels of residues which do not present any problems to the consumer. As with other commodities, isolated instances of residue levels that violate the regulations do occur. These justify the value of monitoring programs.

Whether the situation is the same for free-range or organic eggs remains to be determined. As with milk and some other foods, no comparable figures for free-range or organic eggs are available.

Very little information has been published recently on chemical residues in eggs in the scientific literature and old data are probably not relevant, given the changes that have been made in curtailing the use of chemicals in modern poultry production. Some regions, however have more definite problems from chemical residues, resulting from a high concentration of farms in heavily populated areas close to factories and industrial complexes. Parts of Europe have dioxin levels in the soil that are of concern. As explained in an earlier chapter, low levels of dioxins are common in the environment. They are released through natural processes such as forest fires and volcanic eruptions, and through industrial processes such as combustion or incineration of industrial waste or chemical manufacturing. These compounds can remain in the environment for decades. They accumulate in the fatty tissues of food animals and in fatty tissue in the human body. Studies have shown that prolonged exposure to elevated levels of dioxin may have long-term adverse health effects. Animal feed contaminated with dioxins was reported in Ireland in late 2008, causing the authorities to take swift and effective remedial action.

Poultry and livestock allowed access to pasture may be at a greater risk of accumulating dioxins and similar chemicals in their tissues. This has implications for organic producers who allow their stock to range outdoors. Dutch researchers have reviewed European findings on dioxins in organic eggs (DeVries *et al.*, 2006). They found that organic eggs from farms in The Netherlands and other EU countries contained more dioxin than conventional eggs. High levels were found mostly on smaller farms, possibly because the chickens made more extensive use of outdoor runs. In addition, a significant number of organic farms produced eggs with a dioxin content that exceeded the EU standard. The increased content of dioxin in these

Table 8.2   Canadian compliance summary for eggs and egg products, 5-year study of chemical residues

	2003/2004		2002/2003		2001/2002		2000/2001		1999/2000	
	n	% OK	*n*	% OK	*n*	% OK	*n*	% OK	*n*	% OK
Domestic product testing										
Antibiotics										
Eggs	1 465	100.00	243	100.00	118	100.00	95	100.00	103	100.00
Egg products	0	–	0	–	0	–	0	–	0	–
Subtotal	1 465		243		118		95		103	
Chloramphenicol										
Eggs	68	100.00	37	100.00	13	100.00	76	100.00	103	100.00
Egg products	4	100.00	0	–	12	100.00	2	100.00	22	100.00
Subtotal	72		37		25		78		125	
Clopidol										
Eggs[1]	260	98.83	245	99.59	153	98.69	50	100.00	61	100.00
Egg products	17	100.00	0	–	0	–	0	–	0	–
Subtotal	274		245		153		50		61	
Coccidiostats (multi-residue)										
Eggs	313	100.00	261	100.00	205	100.00	2	100.00	28	100.00
Egg products	16	100.00	0	–	0	–	0	–	0	–
Subtotal	329		261		205		2		28	
Fluoroquinolones										
Eggs	377	100.00	281	100.00	208	100.00	0	–	60	100.00
Egg products	17	100.00	0	–	0	–	0	–	0	–
Subtotal	394		281		208		0		60	
Halofuginone										
Eggs	400	100.00	38	100.00	13	100.00	48	100.00	96	100.00
Ionophore (multi-residue)										
Eggs	399	99.50	30	100.00	0	–	0	–	–	–
Sulfonamides										
Eggs	357	100.00	37	100.00	13	100.00	60	100.00	103	100.00
Egg products	0	–	0	–	12	100.00	1	100.00		
Subtotal	357		37		25		61			
Mycotoxins										
Eggs	0	0	0	–	0	–	0	–	–	–
Pesticides										
Eggs	323	100.00	287	100.00	221	100.00	0	–	63	100.00
Egg products	17	100.00	0	–	7	100.00	0	–	41	100.00
Diquat pretest	0	–	0	–	0	–	0	–	7	–
Pesticide pretest	0	–	0	–	0	–	3	–		
Subtotal	340		287		228		3	–	111	
Domestic total	*5 498*		*1 459*		*975*		*337*		*687*	

[1] Includes three suspect samples not included in the compliance calculation.

(*Continued*)

Table 8.2 (*Continued*)

	2003/2004		2002/2003		2001/2002		2000/2001		1999/2000	
	n	% OK	*n*	% OK	*n*	% OK	*n*	% OK	*n*	% OK
Imported eggs and egg products										
Antibiotics	162	100.00	126	100.00	88	100.00	10	100.00	98	100.00
Chloramphenicol	17	100.00	43	100.00	8	100.00	54	100.00	107	100.00
Clopidol	180	100.00	170	100.00	95	100.00	63	100.00	61	100.00
Coccidiostats	17	100.00	44	100.00	5	100.00	44	100.00	18	100.00
Decoquinate	216	100.00	240	100.00	150	100.00	53	100.00	0	–
Fluoroquinolones	258	100.00	213	100.00	145	100.00	0	–	36	100.00
Halofuginone	12	100.00	43	100.00	1	100.00	50	100.00	82	100.00
Ionophores	12	100.00	43	100.00	0	–	0	–	0	–
Sulfa drugs	12	100.00	43	100.00	1	100.00	19	100.00	94	100.00
Mycotoxins	0	–	0	–	0	–	0	–	44	100.00
Pesticides incl. PCBs	222	100.00	190	100.00	204	100.00	0	–	94	100

(*Source*: CFIA (2006). 2005–2006 National Chemical Residue Monitoring Program Annual Report, Canadian Food Inspection Agency. Reproduced with the permission of the Minister of Public Works and Government Services, Canada, 2010.)

eggs was attributed to a greater intake of dioxins from environmental sources, including plants, feed, soil, worms and insects. Consumption of worms and insects and particularly ingestion of soil are considered to be important causes of high dioxin levels in eggs. Compared with hens on conventional and free-range farms, organic hens are more exposed to these sources because of easier access to outdoor runs. Plants appear to be relatively unimportant as a source of dioxins. Also commercial organic feeds generally have very low dioxin contents, but not much is known about non-commercial feed.

The situation in parts of Europe is serious enough that several studies had to be undertaken to find a solution to the problem of chemical contamination. Countries with a long history of intensive agriculture have soils that are likely to contaminate eggs. Belgium is one such country. The chemical contamination of free-range eggs in Belgium was studied by Overmeire *et al.* (2006). Included were manganese, cobalt, nickel, copper, zinc, arsenic, selenium, molybdenum, cadmium, antimony, thallium, lead and mercury, and selected persistent organochlorine compounds (dioxins, and dioxin-like polychlorinated biphenyls (PCBs), dichlorodiphenyltrichloroethane (DDT) and metabolites as well as other chlorinated pesticides). The eggs tested were free range obtained from private and commercial farms.

The study found that eggs from private farms were more contaminated than eggs from commercial farms. The ratios of levels in eggs from private farms to commercial farms ranged from 2 to 8 for the toxic contaminants lead, mercury, thallium, dioxins, PCBs and the group of DDT compounds.

It was calculated that an extensive consumption of eggs from private farms in Belgium was likely to result in an intake of toxins that would exceed the tolerable intake. The researchers also postulated that environmental pollution was the cause of the higher contamination of eggs from private farms.

Other research in Belgium confirmed the contamination of eggs from free-range chicken with dioxins and dioxin-like PCBs (Schoeters and Hoogenboom, 2006). Eggs from free-range hens owned by private owners in the districts north of Antwerp were found to contain high levels of polychlorinated dibenzodioxins (PCDD) (9.9 pg WHO-TEQ per gram of fat; Pussemier et al., 2004). As the Belgian researchers pointed out, dioxins and dioxin-like (DL) PCBs are persistent organic pollutants that enter the body mainly via the feed or food. A small margin exists between current exposure levels in humans and the levels causing biological effects. Therefore, stringent control of concentrations of these contaminants in food and feed is needed. Eggs from free-range chickens are increasingly becoming an important part of the diet. These eggs have a higher risk of being contaminated with increased levels of dioxins and DL-PCBs than barn or cage eggs. Intake of soil particles from environmentally contaminated areas may contribute to elevated dioxin levels in free-range chicken eggs. The data showed that current soil levels of dioxins and DL-PCBs in residential and agricultural areas in Europe are frequently too high for the production of free-range eggs with dioxin levels below the current allowable limits in the EU.

None of the above testing has been applied to organic eggs. A general assumption is that residues might be lower in eggs from hens allowed to range outdoors, since fewer or no medication is permitted. However, increased contact with soil, bedding and other materials in the environment, rodents and other pests, and toxin-producing fungi, etc., may lead to increased residues from these sources in eggs. Treatment of roosts for mite control might also result in residues in eggs.

Chemical residues in poultry and eggs produced in free-range or organic systems were reported in a detailed study conducted in The Netherlands (Kan, 2005). This review found that organic production appeared to add an extra risk for dioxin contamination of eggs, based on European data. It was found that poultry in outdoor farming systems in that part of Europe are more exposed to infectious agents and chemical contaminants than those housed in barns. Dioxin levels were higher in eggs from some organic systems than from conventional systems. Both general and local environmental pollution with dioxins were probably responsible for this.

The review by Kan found that high levels of dioxins in eggs from organic systems have been reported from several European countries (Belgium, Germany, France, Ireland, Sweden, Switzerland and The Netherlands). Eggs from other outdoor systems seemed to contain lower levels. No explanation for this difference was available, suggesting that feed might be the source. The higher contamination in organic systems seemed also to be related to soil contamination and ingestion of soil by the hens. Penta-chlorophenol (PCP) treated wood was also a source of dioxin contamination of farm animals. Drug residues were generally not higher in systems using deep litter, but, due to recirculation via the litter, persistence of residues was often considerably longer.

The conclusion of the above findings is that the environment in parts of Europe may be contaminated by chemicals. Some of these are persistent and it may take time and continued effort to improve the situation. Organic poultry production may not be suitable for the regions that are most heavily contaminated, and egg producers should not attempt to produce organic or free-range eggs there.

Egg quality

Research findings

In the US and in other countries the standards used for grading shell eggs have been developed on the basis of both interior and exterior quality. Commercially, eggs are graded simultaneously for exterior and interior quality. When determining the grade of an egg, the factor with the lowest grading value determines the overall grade of the egg.

In the United States, egg grades include AA quality, A quality, B quality, and dirty. Only AA and A quality eggs are sold in supermarkets. US Grade AA eggs have whites that are thick and firm; yolks that are high, round, and practically free from defects; and clean, unbroken shells. US Grade A eggs have characteristics of Grade AA eggs except that the whites are "reasonably" firm. This is the quality most often sold in stores. Grade AA and Grade A eggs are best for frying and poaching where appearance is important, and for any other uses. US Grade B eggs have whites that may be thinner and yolks that may be wider and flatter than eggs of higher grades. The shells must be unbroken, but may show slight stains. This quality is seldom found in retail stores because they are generally used to make liquid, frozen, and dried egg products.

The European grading system is more detailed, with Grade A eggs required to have the following minimum characteristics:

> shell and cuticle to be normal, clean, undamaged; the air space height not to exceed 6 mm and stationary (for eggs to be marketed as "extra" it may not exceed 4 mm; the white to be clear, limpid, of gelatinous consistency, and free of extraneous matter of any kind; the yolk to be visible on candling as a shadow only without clearly discernible outline, not moving appreciably away from the centre of the egg on rotation, and free of extraneous matter of any kind; the germ cell to show imperceptible development; and freedom from extraneous

odors. The regulations further require that Grade A eggs shall not be washed, or cleaned by any other means, before or after grading. Accordingly, eggs washed in accordance with Article 6(4) of Regulation (EEC) No. 1907/90, even where they fulfil the criteria applicable to Grade A eggs, may not be marketed as Grade A eggs and must be marked "washed eggs". Also, Grade A eggs shall not be treated for preservation or chilled in premises or plants where the temperature is artificially maintained at less than 5 °C. However, eggs which have been kept at a temperature below 5 °C during transport of not more than 24 hours or on retail premises or in annexes thereto shall not be considered as chilled, provided the quantity stored in these annexes does not exceed the requirements for three days of retail sale on the premises in question. Accordingly, eggs chilled in accordance with Article 6(5) of Regulation (EEC) No. 1907/90 may not be marketed as Grade A eggs, even where they fulfil the criteria applicable to Grade A eggs. They shall be marketed as "chilled eggs". Grade B eggs are eggs which do not meet the requirements applicable to eggs in Grade A. They may be passed only to food industry undertakings approved in accordance with Article 6 of Directive 89/437/EEC or to non-food industry undertakings.

The main difference between these major markets is that Grade A eggs in the US can be washed (and usually are) whereas in Europe the Grade A eggs are required to be unwashed. Managing layer flocks in European farms to deliver clean eggs is therefore very important.

The other main aspect of grading is shell quality. In this case the most important requirement is that the shell is sound and free from cracks. The color is not taken into account in grading, though some markets prefer white-shelled eggs and others prefer brown-shelled eggs. Generally, darker-colored brown eggs have thicker shells which are more resistant to breaking, because the hens that lay them lay fewer eggs and so have more calcium to deposit in each shell. However, shell color does not influence the internal quality of the egg such as flavor, nutritional composition, or cooking characteristics. Shell color is derived from pigments in the outer layer of the shell. It is primarily a breed characteristic, breeds with white earlobes generally laying white eggs while breeds with red earlobes generally lay brown eggs.

Yolk color varies, being dependent on pigmenting agents in the feed. Birds that have access to green plants or have alfalfa or yellow corn in the diet produce eggs with yolks that are more highly pigmented (yellow to orange). Some consumers like pale yolks, others orange-colored yolks. The eggs from organically fed (and free-range) hens are usually highly pigmented because of their access to grass and other forages.

A limited amount of research has been done on how organic production affects egg quality, related mainly to lipid components. In their authoritative book, *Egg Science and Technology*, Stadelman *et al.* (1995) concluded that:

> ..."organic" eggs may be promoted as much safer and more nutritious than those produced by hens on the usual commercial rations. Since pesticides, fungicides, herbicides, and commercial fertilizers must be proved safe before they can be used in the prescribed amounts at specific times in crop production, there is no reason for concern about eating eggs produced by hens on commercial rations. Organic eggs are no higher in nutritive value than regular eggs. If the ration for hens on an organic diet is not so well balanced as the usual commercial laying ration, the nutritive value of organic eggs tends to be lower.

The point made about the nutritional balance and how it affects overall egg composition was verified in a subsequent report by European researchers (Minelli *et al.*, 2007). In this study over 1400 eggs from organic and conventional egg farms were tested. The same strain of bird was used on both types of farm. The data (Table 8.3) showed that eggs from the organic farms were smaller (64.4 vs. 66.2 g), with yolk, albumen and eggshell weights being significantly lower compared with eggs from the farms using conventional,

Table 8.3 Characteristics of eggs produced by hens under conventional (battery cages) or an organic system

	Housing system		
	Battery cage	Organic	Significance of difference $P < 0.01$
Egg wt (g)	66.2	64.4	$P < 0.01$
Yolk wt (g)	16.7	15.8	$P < 0.01$
Egg shell wt (g)	6.21	6.11	$P < 0.01$
Albumen wt (g)	43.3	42.4	$P < 0.01$
Yolk/albumen (%)	0.39	0.38	$P < 0.01$
Shell/egg (%)	9.51	9.42	NS
Breaking force (kg)	3.265	3.135	$P < 0.01$
Haugh units	70.9	78.6	$P < 0.01$
Color number	13.02	8.28	$P < 0.01$
Dry matter (%)	50.7	51.1	NS
Lipid (%)	31.0	31.0	NS
Protein (%)	16.7	17.1	$P < 0.01$
Ash (%)	1.63	1.61	NS
Cholesterol (%)	1.21	1.26	$P < 0.01$

(*Source*: reproduced from Minelli, G., Sirri, F., Folegatti, E., *et al.* (2007). Egg quality traits of laying hens reared in organic and conventional systems. *Italian Journal of Animal Science* **6**, 728–730, with permission of ASPA, Italy.)

caged systems. This resulted in the yolk/albumen ratio being lower in the organic eggs (0.38 vs. 0.39). The percentage of eggshell was not affected by the housing system, but eggshell strength was higher in the eggs produced in the conventional system (3.265 vs. 3.135 kg). Organic eggs showed significantly higher contents of protein (17.1 vs. 16.7 percent) and cholesterol (1.26 vs. 1.21 percent). The yolks of organic eggs were paler than the yolks of conventional hens, attributed to a ban on the use of pigmenting agents in the organic diets and a lack of access to grass outdoors because of an outbreak of avian influenza. No details were provided on the composition of the diets used or of the total eggs laid by each group.

The lower nutritive value of the organic eggs was attributed to the feeding regimen, which did not support the birds with adequate levels of micronutrients needed to maximize their production. This effect, which resulted from the restrictions placed on feed composition in organic poultry (and pig) production, is obvious from observations of the industry, and is expected to become more severe when all organic feed in Europe must be 100 percent organic by 2012. Until that date some non-organic ingredients are permitted in the feed, which has helped to minimize the problem. The main deficiency in organic feed used in poultry production in Europe is a shortage of suitable feedstuffs and a dietary deficiency of methionine exacerbated by a ban on the use of solvent-extracted soybean meal and the pure (non-synthetic) form of this amino acid.

Cherian *et al.* (2002) also reported a significant reduction in the percentage of yolk and a corresponding increase in the percentage of white in eggs from hens fed an organic diet.

This feature of layer feed not being required in Europe to be 100 percent organic until 2012 makes interpretation of some research reports quite difficult. Researchers who purchase eggs at a retail outlet for testing have no control over the composition of the diets fed to the birds. The only information available to these researchers is the label information. Consequently, the most useful research reports are those in which the feed composition was controlled in the research study and the composition of the diet stated.

Another difficulty facing researchers who purchase different types of eggs (and other foods) for testing is that each type should be compared at the same degree of freshness. If not, the effects of production method and age of product (which affects freshness) can be confounded. This feature was demonstrated in a research study by Hidalgo *et al.* (2008), who compared market eggs from four housing systems, i.e. cage, barn, free-range and organic. Organic eggs had the highest whipping capacity and foam consistency but the lowest freshness (the highest air-cell height) and lowest albumen quality (Haugh unit). Cage eggs had the lowest whipping capacity but the highest shell resistance to breaking. The variables with the most discriminant power were shell-breaking resistance, whipping capacity, egg protein content, and shell thickness. It was possible to use these to separate cage eggs from the others, but not from each other. Obviously the data would have been more valuable if all of the eggs had been at the same stage of freshness.

Cherian *et al.* (2002) compared the quality of branded large eggs purchased from retail stores in the US. The egg types were SP1, from hens fed a diet free of animal fat; SP2, certified organic free-range brown eggs; SP3, non-caged unmedicated brown eggs; SP4, non-cage vegetarian diet brown eggs; SP5, naturally nested non-caged. Regular white-shelled eggs were the control. A significant difference was observed in the egg components. The percentage of yolk was lower in SP2 and SP4 with a corresponding increase in the percentage of white. The percentage of shell was lower in SP4 and SP5. The total edible portion was greater in SP4 and SP5. The yolk/white ratio was greater in SP3 eggs. The total lipid content was lower in SP4 eggs.

Much of the research with laying hens has been directed at the fatty acid composition of the egg, in both conventional and organic feeding systems. For instance Milinsk *et al.* (2003) fed four diets with different fatty acid composition to laying hens and reported a change in the relative amounts of saturated and unsaturated fatty acid in yolk lipids but not in cholesterol or lipid contents of the egg yolk. The major effects of the diets were observed in fatty acids C16:0, C18:0, C18:1n9, C18:2n6, C20:4n6, C20:5n3 and C22:6n3. The addition of oils to the diets fed to hens resulted in the production of eggs with higher n3/n6 and PUFA/SFA fatty acid ratios than in the eggs from control hens.

Results on this issue are not, however, consistent. Poupoulis *et al.* (2009) determined the fatty acid composition of egg yolk of hens from farms using conventional, organic, free-range and omega-3 feeding systems. Results showed that the farming system significantly influenced the PUFA content, total omega-3 and omega-6 fatty acids as well as the ratios of PUFA/SFA and omega-6/omega-3 fatty acids. These researchers concluded that eggs produced on omega-3 farms had the best fatty acid profile, followed by organic farms. Samman *et al.* (2009) studied the fatty acid composition of commercially available conventional, certified organic, and omega-3 eggs. Organic egg yolk contained a higher percentage of palmitic and stearic acids than did conventional yolk, with no differences observed in the monounsaturated or polyunsaturated fatty acid compositions. Compared with organic and conventional eggs, omega-3 egg yolk contained lower percentages of myristic and palmitic acids, and higher omega-3 fatty acids. However, the researchers concluded that the small differences in saturated fatty acids observed in the present study are unlikely to have any significant metabolic effect on the consumer.

Some research studies have addressed the issue of whether scientific methods could be used to identify organic eggs. For instance, Rogers (2009) tested stable isotopes of carbon and nitrogen in market eggs laid by hens from cage, barn, free-range, and organic systems. In general, free-range and organic eggs had up to 4 percent higher contents of ^{15}N than eggs from hens in cages or housed in barns. One sample of free-range and two samples of organic eggs had ^{15}N values within the range of caged or barn-laid eggs, suggesting either that these eggs were mislabeled (the hens were raised under "battery" or "barn" conditions, and not permitted to forage outside) or that there was insufficient animal protein gained by foraging to shift the ^{15}N values of their primary food source. ^{13}C did not appear to be of similar potential value as the stable nitrogen isotopes as an authentication tool for the egg industry.

Consumer findings

Several studies have been conducted in Europe to investigate consumer perceptions of the egg production industry, e.g. Mirabito and Magdelaine (2001). In the French study more than 95 percent of those interviewed stated that freshness and safety were the main criteria for egg purchase. Packaging, quality and brand were also important. Generally, the ideal production system was considered to be one based on relatively few hens, natural feed, and freedom of movement. Eighty-five percent of respondents were of the opinion that free-range systems resulted in fresh (or safe) eggs compared with 27 percent for battery systems. Ninety-five percent of respondents thought that keeping laying hens outside was the best system to improve bird welfare. Respondents were divided on concerns over bird welfare and their willingness to pay extra for free-range eggs. Eighteen percent were unconcerned about bird welfare and were not willing to pay extra for free-range eggs; 39 percent were concerned and prepared to pay 0 to 50 percent more; 27 percent were very concerned and were already buying organic products and willing to pay 50 percent more for the eggs.

A UK study showed that consumers bought organic eggs because they were perceived as being healthier and free of chemicals and genetically modified materials, and because they tasted better (Stopes et al., 2001). In addition, consumers expected that the laying flocks were maintained under more humane and improved welfare conditions.

The above studies suggest several important conclusions. First:, organic poultry meat and eggs should be produced in such a way that they meet the expectations of consumers both before and after purchase. Second, the willingness of consumers to pay a premium for organic produce is not unlimited. These conclusions indicate that organic poultry producers need to strive to produce a high-quality product as economically as possible.

Conclusions

There is no evidence that the cholesterol content of eggs is lower in organic than in regular eggs, although the fatty acid profile may be more favorable.

Also, contamination with *Salmonella* and other food-poisoning organisms is not consistently different between organic and regular eggs, and may be higher in organic eggs. Antibiotic residues in eggs are no longer a concern for the consumer. All commercial poultry – free-range, barn or caged – are fed diets free of antibiotics, unless administered by a veterinarian and, if the hens are treated, the eggs are not sent to market. Little or no contamination of regular eggs with chemical residues has been reported, although contaminated soils in parts of Europe appear to cause residues of dioxins in the eggs of hens allowed to range outdoors. Consumers need to take note of these findings when opting to buy eggs from hens housed on floor-based or organic systems.

Most of the research on nutritional composition of organic eggs has been directed at the fatty acid composition. Some findings indicate that the PUFA content is enhanced by organic feeding, but the effect is not consistent and can be achieved by the use of appropriate dietary ingredients. In other respects the nutritional composition of organic and conventional eggs is similar.

References

Behrenbeck, T. (2008). Are chicken eggs good or bad for my cholesterol? http://www.mayoclinic.com/health/cholesterol/HQ00608, accessed July 23, 2009.

Berg, C. (2002). Health and welfare in organic poultry production. *Acta Veterinaria Scandinavica* **43**, S37–S45.

CDC (2010). Investigation Update: Multistate Outbreak of Human *Salmonella* Enteritidis Infections Associated with Shell Eggs. US Centers for Disease Control and Prevention, Atlanta, Georgia. http://cdc.gov/salmonella/enteritidis/, accessed December 20, 2010.

CFIA (2006). 2005–2006 National Chemical Residue Monitoring Program Annual Report, Canadian Food Inspection Agency. http://www.inspection.gc.ca/english/fssa/microchem/resid/2005-2006/annue.shtml#ct4_1_2, accessed October 1, 2009.

Chapman, B. (2008). Raw Eggs in Mayonnaise Blamed as *Salmonella* Infects 18 on Guernsey Island in the English Channel. International Food Safety Network. http://www.foodsafety.ksu.edu/articles/1208/iFSN-infosheet-6-26-08.pdf, accessed July 24, 2008.

Cherian, G., Holsonbake, T.B. and Goeger, M.P. (2002). Fatty acid composition and egg components of specialty eggs. *Poultry Science* **81**, 30–33.

DAFF (2010). Australian National Residue Survey 2008–2009. www.daff.gov.au/agriculture-food/nrs/animal2, accessed October 25, 2010.

De Reu, K., Rodenburg, T.B., Grijspeerdt, K. *et al.* (2009). Bacteriological contamination, dirt, and cracks of eggshells in furnished cages and noncage systems for laying hens: an international on-farm comparison. *Poultry Science* **88**, 2442–2448.

DeVries, M., Kwakkel, R.P. and Kijlstra, A. (2006). Dioxins in organic eggs: a review. *NJAS - Wageningen Journal of Life Sciences* **54**, 207–221.

Foods Standards Agency (2004). Report of the Survey of Salmonella Contamination of UK Produced Shell Eggs on Retail Sale. UK Foods Standards Agency, London. http://www.food.gov.uk/multimedia/pdfs/fsis5004report.pdf, accessed April 10, 2010.

Fossum, O., Jansson, D.S., Etterlin, P.E. and Vågsholm, I. (2009). Causes of mortality in laying hens in different housing systems in 2001 to 2004. *Acta Veterinaria Scandinavica* **51**, 3.

Gray, J. and Griffin, B. (2009). Eggs and dietary cholesterol – dispelling the myth. *Nutrition Bulletin* **34**, 66–70.

Greene, C.M., Waters, D., Clark, R.M., *et al.* (2006). Plasma LDL and HDL characteristics and carotenoid content are positively influenced by egg consumption in an elderly population. *Nutrition and Metabolism* **3**, 6.

Harman, N.L., Leeds A.R. and Griffin, B.A. (2008). Increased dietary cholesterol does not increase plasma low density lipoprotein when accompanied by an energy-restricted diet and weight loss. *European Journal of Nutrition* **47**, 287–293.

Hidalgo, A., Rossia, M., Clericia, F. and Rattia, S. (2008). A market study on the quality characteristics of eggs from different housing systems. *Food Chemistry* **106**, 1031–1038.

Holub, B.J. (2002). Review. Clinical nutrition: 4. Omega-3 fatty acids in cardiovascular care. *Canadian Medical Association Journal* **166**, 608–615.

Hu, F.B., Stampfer, M.J. and Rimm, E.B., *et al.* (1999). A prospective study of egg consumption and risk of cardiovascular disease in men and women. *Journal of the American Medical Association* **281**, 1387–1394.

Ireland National Food Residue Database (2009). Residue Studies Group, Food Safety Department, Ashtown Food Research Centre, Teagasc, Ireland. http://nfrd.teagasc.ie, accessed October 11, 2010.

Kan, C.A. (2005). Chemical residues in poultry and eggs produced in free-range or organic systems. XVIIth European Symposium on the Quality of Poultry Meat and XVIIth European Symposium on the Quality of Eggs and Egg Products, 28–36.

Kritchevsky, S.B. and Kritchevsky, D. (2000). Egg consumption and coronary heart disease: An epidemiologic overview. *Journal of the American College of Nutrition* **19**, 549S–555S.

Miele, M. and Parisi, V. (1999). The nature of consumer concerns about animal welfare and the impact on food choice – the Italian Focus Groups Report. Report for EU FAIR Project CT98-3678. Università Degli Studi de Pisa, Italy.

Milinsk, M.C., Murakami, A.E., M. Gomes, S.T.M., *et al.* (2003). Fatty acid profile of egg yolk lipids from hens fed diets rich in n-3 fatty acids. *Food Chemistry* **3**, 287–292.

Minelli, G., Sirri, F., Folegatti, E., *et al.* (2007). Egg quality traits of laying hens reared in organic and conventional systems. *Italian Journal of Animal Science* **6**, 728–730.

Mirabito, L. and Magdelaine, P. (2001). Effect of perceptions of egg production systems on consumer demands and their willingness to pay. *Sciences et Techniques Avicoles* **34**, 5–16.

Overmeire, I. van, Pussemier, L., Waegeneers, N. *et al.* (2009). Assessment of the chemical contamination in home-produced eggs in Belgium: general overview of the CONTEGG study. *Science of the Total Environment* **407**, 4403–4410.

Poupoulis, C., Salepi, M., Skapetas, B. and Touska, M. (2009). Fatty acid composition of hens' egg yolk from conventional, organic, Ω-3 and free range farming types. *Epitheorēsē Zootehnikēs Epistēmēs* **40**, 73–84.

Pussemier, L., Mohimont, L., Huyghebaert, A. and Goeyens, L. (2004). Enhanced levels of dioxins in eggs from free range hens; a fast evaluation approach. *Talanta* **63**, 1273–1276.

Rodenburg, T.B., Van Der Hulst-Van Arkel, M.C. and Kwakkel, R.P. (2004). Campylobacter and Salmonella infections on organic broiler farms. *NJAS–Wageningen Journal of Life Sciences* **52**, 101–108.

Rogers, K.M. (2009). Stable isotopes as a tool to differentiate eggs laid by caged, barn, free range, and organic hens. *Journal of Agricultural and Food Chemistry* **57**, 4236–4242.

Samman, S., Kung, F.P., Carter, L.M. *et al.* (2009). Fatty acid composition of certified organic, conventional and omega-3 eggs. *Food Chemistry* **116**, 911–914.

Schoeters, G. and Hoogenboom, R. (2006). Contamination of free-range chicken eggs with dioxins and dioxin-like polychlorinated biphenyls. *Molecular Nutrition and Food Research* **50**, 908–914.

Simopoulos, A.P. (2009). Essential fatty acids in health and chronic disease. *American Journal of Clinical Nutrition*, **70**, 560S–569S.

Stadelman, W.J., Newkirk, D. and Newby, L. (1995). *Egg Science and Technology*, Fourth edition. CRC Press, Boca Raton, Florida.

Stopes, C., Duxbury, R. and Graham, R. (2001). Organic egg production: consumer perceptions. In: Younie, D. and Wilkinson, J.M. (eds). Proceedings of a Conference on Organic Livestock Farming, Heriot-Watt University, Edinburgh, and University of Reading, February 9 and 10, 2001, pp. 177–179.

Sulonen, J., Kärenlampi, R., Holma, U. and Hänninen, M.-L. (2007). Campylobacter in Finnish organic laying hens in Autumn 2003 and Spring 2004. *Poultry Science* **86**, 1223–1228.

Trulson, M.F., Clancy, R.E., Jessop, W.J.E., *et al.* (1990). Comparisons of siblings in Boston and Ireland: physical, biochemical, and dietary findings. *Nutrition Reviews* **48**, 370–372.

US Food and Drug Administration (2009). www.fda.gov/Safety/Recalls/ArchiveRecalls/2009/ucm128502.htm, accessed February 14, 2011.

USDA (2010). US Department of Agriculture Pesticide Data Program. http://www.ams.usda.gov/pdp, accessed April 26, 2010.

Van Immerseel, F., Cauwerts, K., Devriese, L.A., *et al.* (2002). Feed additives to control Salmonella in poultry. *World's Poultry Science Journal* **58**, 501–513.

9 Is Organic Food Safer?

As outlined in Chapter 2, one of the main reasons for consumers opting for organically produced food is the perception that conventionally produced food contains harmful residues of dangerous chemicals and pesticides, hormones and so on. As a result organic food is viewed as being safer and more "healthful". It is important now to review the documented evidence outlined in Chapters 3 to 8 and determine the extent to which this perception can be justified.

Residues

The documented findings on residues of pesticides and chemicals in conventionally produced foods of domestic origin indicate that these cannot be considered a major concern. It is true that, in general, conventionally produced food may contain higher concentrations of pesticide residues than organically produced food (e.g. Pussemier et al., 2006), but the levels in question are much lower than the levels that can be considered deleterious. Some of the residues were detected merely at the level of detection.

Another point that is clear from the monitoring data published by several countries is that the contamination rate of conventional foods is falling.

A comparison of organic and conventional foods would be much clearer if organic foods were subjected to the amount of official testing required for conventional foods.

Vegetable produce

Based on the documented evidence, levels of pesticides are likely to be lower in organic produce than in regular produce, but a reduced intake in our diet is probably not meaningful in terms of human health. Food-poisoning germs are a lot more important and can occur both in organic and regular foods.

Organic Production and Food Quality, First Edition. Robert Blair.
© 2012 John Wiley & Sons, Ltd. Published 2012 by John Wiley & Sons, Ltd.

Fruit

Although it seems logical that lower levels of chemical pesticide residues should be found in organic foods, there is little documented evidence in the scientific literature or in the results of governmental food monitoring programs to prove this point convincingly. Also, the available findings indicate that the low levels of residues reported in conventional foods rarely exceed acceptable limits.

Cereal grains

Based on the documented evidence, it does not appear that chemical and pesticide residues in conventional cereals are a major health concern for consumers. A more definite area of concern is the possibility of mycotoxin contamination, both in conventional and organic cereals.

Meat

Concerns about "mad-cow disease" and about possible hormone, chemical and antibiotic dangers in regular meats are not justified by the known facts. Food regulations in developed countries are now effective in addressing these issues. Incorrect handling of meat, especially poultry meat, that leads to contamination with food-poisoning organisms is a recognized threat to the consumer and is of greater importance than whether the meat is from animals raised conventionally or organically.

Monitoring studies conducted by official agencies indicate that the meat supply in North America is not heavily contaminated by chemical and pesticide residues or dioxins and similar environmental contaminants. The levels found in various surveys can be regarded as trace amounts. A few instances of violative residue levels have been reported in meat in several countries but all of the published comparisons of organic and conventional meats do not indicate any major difference in contamination rate. Residues of chlorinated-hydrocarbon and organophosphate pesticides have been reported both in organic and conventionally raised cattle. There are insufficient data to allow an accurate comparison to be drawn. The limited findings available on organic meat indicate that it is highly unlikely there is any difference in occurrence of chemical residues of drugs, vaccines, pesticides, antibiotics and/or growth promotants in conventional and organic beef. Organic meat is not completely devoid of residues.

The United States, Canada and Australia are among countries that permit the use of hormone implants in cattle. Monitoring studies have proved their safety, deeming them safe and presenting no harm to the consumer of meat from treated cattle. Hormones are not used in pig or poultry production.

In general the concentrations of contaminants have been found to be very low in both organic (farmed) and wild (conventional) fish.

Milk

The quality of regular milk is extremely well controlled in Western countries and is of high quality, whether from cows treated with bovine somatotropin or not. Reviews of the published findings by medical, nutritional and animal science experts have concluded that milk from rbST-supplemented cows has been a part of the US food supply since receiving FDA approval over 15 years ago and its use has not been associated with any scientifically documented detrimental effects on human health. The available evidence indicates that regular and organic milk contain similar trace levels of chemical and pesticide residues.

Milk safety programs now require that all bulk milk tankers be sampled and analyzed for animal drug residues and antibiotics before the milk is processed. Any tanker found positive is rejected for human consumption.

Data from several countries indicate a 100 percent compliance rate for domestically produced milk in terms of pesticide and chemical residues. The data from official monitoring agencies do not include comparable figures for organic dairy products.

Eggs

There is no evidence that the cholesterol content of eggs is lower in organic than in regular eggs, although the fatty acid profile may be more favorable. Also, contamination with *Salmonella* and other food-poisoning organisms is not consistently different between organic and regular eggs, and may be higher in organic eggs. Antibiotic residues in eggs are no longer a concern for the consumer. All commercial poultry – free-range, barn or caged – is fed diets free of antibiotics, unless administered by a veterinarian and, if the hens are treated, the eggs are not sent to market. Little or no contamination of regular eggs with chemical residues has been reported, although contaminated soils in parts of Europe appear to cause residues of dioxins in the eggs of hens allowed to range outdoors.

The above conclusions do not necessarily apply to all other countries. For instance, parts of Europe have soils that are contaminated with residues of pesticides such as dioxins. Foods grown in such areas, conventional and organic, are likely to be contaminated for some time.

Residues of heavy metals have been reported in fruit juice, both organic and conventional. The copper level has been reported to be higher in some organic fruit and produce but this does not appear to be a problem, especially since copper-based fungicides have now been banned.

Food poisoning

It may well be that organic farms are at a greater risk than conventional farms from food poisoning, because of the use of animal manure as fertilizer and a greater exposure of stock to wild animals and birds. The serious outbreak of *E. coli* sickness in Europe in 2011 highlighted the potential for organic produce to be a source of infection, with the European Commission through its Rapid Alert System for Food and Feed notifying the 27 Member States about one of the confirmed sources responsible for the STEC (shigatoxin-producing *E. coli*) outbreak affecting primarily Germany and four other Member States. This notification followed the reporting by the German authorities that they had identified organic cucumbers from two provinces of Spain (Almeria and Malaga) as one of the sources (EU, 2011) (http://europa.eu/rapid/pressReleasesAction.do?reference=IP/11/653&type=HTML). A subsequent joint technical report (ECDC-EFSA, 2011a), issued by the European Centre for Disease Control and the European Food Safety Authority on June 10, 2011, announced that contaminated bean sprouts from an organic farm in Northern Germany were most likely the source of the outbreak. By then 29 people had died and almost 3000 people had been sickened by the infection in Europe. A subsequent announcement on June 29, 2011 by EFSA-ECDC (2011b) further implicated the consumption of raw fenugreek sprouts as a possible source of the outbreak in France as well as Germany. The ongoing investigation includes a study of the source of the seed as well as the system of sprout production. On July 5, 2011, the European Union voted to ban imports of some seeds and beans from Egypt until October 31, because of their probable connection with the recent *E. coli* outbreaks in Europe. The ban covers imports of Egyptian seeds and beans for sprouting, including legumes, fenugreek and soya beans.

Consumers need to be aware of the facts relating to this infection, especially since it is now known that washing microbe-infected produce or meat is not effective in avoiding food poisoning. However, there is not enough documented data to allow a determination of whether organic farms are safer, less safe or equally at risk as sources of food poisoning. To minimize the risk of contamination several countries such as the United States and Canada do not allow the use of non-composted manure. Another possible risk factor for organic farms is increased contact with wildlife, rodents, insects and birds, which are potential sources of contamination with fecal pathogens. *Salmonella* and related food-poisoning organisms are best controlled in organic and conventional poultry flocks by vaccinations, a procedure that avoids the use of medication and the possibility of the occurrence of residues in eggs and poultry meat.

An approach being considered in some countries is the use of irradiation to render free of infection any vegetable produce that is to be eaten uncooked. Meanwhile consumers worried about the possibility of eating contaminated produce could minimize the risk by purchasing the produce from a local grower whom they trust.

Mycotoxins

Mycotoxin (mold toxin) contamination of foods such as cereal grains and dried fruits is common worldwide and can occur in both organic and conventional products. It is likely that growing and harvesting conditions have a much greater influence on the likelihood of mycotoxin contamination than the production system (organic vs. conventional), although the recorded data do suggest a higher risk with organic grains (see Chapter 5). An adequate quality control program for grains and dried fruits, etc., as well as appropriate storage conditions prior to sale of the products, will help to minimize any mycotoxin problems. Such a program will also minimize any problems in animals fed these products or bedded on cereal straw, and residues in their meat or milk.

There are recorded instances of risks from mycotoxins in fruit juices, unless precautionary steps are taken. One of these mycotoxins is patulin. More frequent and higher concentrations of patulin have been reported in organic fruit juices and purees than in conventional products. The problem can be avoided by ensuring that the fruit is unaffected by spoilage. In addition it is advisable to pasteurize the juice.

Other anti-nutrients

Potato tubers contain glycoalkaloids, natural compounds produced by the potato to repel insects. They can be detected in potatoes by their green color under the skin. These compounds can be toxic to humans. Some research indicated a higher level in conventionally grown potatoes than in organically grown potatoes, but other research indicated the opposite. In none of these reports were the glycoalkaloids present at concentrations high enough to cause problems to humans. Also, the concentrations varied from year to year.

Nitrate

A main effect in vegetables grown organically is a lower nitrate content, due to the use of manure as fertilizer. The significance of nitrate in the human diet is controversial. High intakes of nitrate from drinking water have been shown to cause methemoglobinemia in infants and it has been suggested that high nitrate intakes may form carcinogenic nitrosamines in the stomach. However, dietary nitrate may also exert a protective effect on the circulatory system. Therefore, it is not certain whether high nitrate concentrations in vegetables should be considered detrimental to adults.

Several scientists have, in fact, suggested that nitrate be classified as a nutrient for humans. For instance Hord *et al.* (2009) in a paper published in the *American Journal of Clinical Nutrition* quantified the nitrate and nitrite concentrations in foods, and then incorporated the values into two hypothetical dietary patterns that emphasized high-nitrate or low-nitrate vegetable and fruit choices based on the Dietary Approaches to Stop Hypertension (DASH) diet. They found that nitrate concentrations in these two patterns varied from 174 to 1222 mg, indicating that the hypothetical high-nitrate DASH diet pattern exceeds the World Health Organization's Acceptable Daily Intake for nitrate by 550% for a 60-kg adult. On this basis these researchers suggested that the data call into question the rationale for recommendations to limit nitrate and nitrite consumption from plant foods, and that a comprehensive re-evaluation of the health effects of food sources of nitrates and nitrites is appropriate. On the basis of their findings they also suggested that the strength of the evidence linking the consumption of nitrate- and nitrite-containing plant foods to beneficial health effects supports the consideration of these compounds as nutrients.

That paper resulted in the publication of an Editorial in the same journal. In it Katan (2009) posed the question, "But what is the evidence that nitrate and nitrite are beneficial?" He pointed out that the evidence is still scant that nitrate in the amount present in vegetables lowers blood pressure. Accordingly he suggested that the indications that dietary nitrate or nitrite reduce cardiovascular disease risk are insufficient to relax the standards for nitrate in drinking water and foods. He went on to make an interesting suggestion: "What we need is a trial in which volunteers are fed matched vegetables with high or low nitrate content."

That call was answered in part by a study conducted in Japan (Sobko *et al.*, 2010). The 25 participants of the study were physically active, healthy volunteers (10 men and 15 women) of mean age 36 years and with a BMI<18.5). The study had a randomized cross-over design with two dietary intervention periods during which the subjects received either a Japanese or control (non-Japanese) diet. The participants ate common Japanese vegetables, identified as the daily source of nitrate, during the study period of 10 days. To avoid the concentration differences in nitrate/nitrite in locally produced foods, the participants were provided with fresh vegetables and staple foods from the same store twice a week during both intervention periods. During the control period, participants were instructed to avoid these vegetables. The individuals participating in the study followed the dietary protocols without a significant weight loss or gain. Daily nitrate intake was calculated. Results showed that the nitrate provided naturally by the Japanese traditional diet was 18.8 mg/kg/bodyweight/day, exceeding the Acceptable Daily Intake by five times (ADI, 3.7 mg/kg/

bodyweight). Plasma and salivary levels of nitrate and nitrite were higher as a result of consuming the Japanese traditional diet. Also, the diastolic blood pressure decreased by an average of 4.5 mm Hg during the test period compared to that with the control diet. Systolic blood pressure was not affected. The authors pointed out that the effect was obtained in normotensive subjects, similar to that seen in the recent studies. They concluded that the results further support the importance of the role of dietary nitrate in relation to blood pressure regulation and suggest one possible explanation for the beneficial aspects of traditional Japanese food.

Significance of the findings in relation to health

It is interesting to note the leading risks to health as ranked by the World Health Organization (WHO). This body has calculated that seven of the top leading risks to health are:

- high blood pressure;
- high cholesterol;
- obesity;
- lack of exercise;
- insufficient vegetables and fruit in the diet;
- alcohol consumption;
- smoking.

There is no listing of pesticide or chemical residues in the above. Does this mean, then, that residues are a lot less important in affecting our health than we might think?

The possible relationship between residues in our food and the incidence of cancer is of great importance and has been studied intensively because of fears that residues might build up in our bodies and cause this disease. It is gratifying, therefore, to note that all of the reviews conducted to date on this topic indicate that fears of residues in the diet being cancer-causing are unfounded. An important statement on the possible relationship between cancer incidence and pesticide/herbicide use in food production was made by the World Cancer Research Fund/American Institute for Cancer Research in 2007. The fact that this report did not find any relationship between herbicide or pesticide residues in food and the incidence of cancer should be reassuring to consumers, and it is surprising that the report received very little media attention when it was presented. Its publication was the culmination of an exhaustive study, involving experts in the cancer field from a large number of countries in Africa, Asia, Australia/New Zealand, Europe and North America.

The report stated that "The Panel considers that the evidence is insufficient to conclude that usual intakes of industrial, agricultural, and other chemicals have an effect on the risk of human cancer" (section 4.7.2). In a section (4.9.1.1) dealing with pesticides and herbicides the Panel stated: "the chlorinated pesticide dichloro-diphenyl-trichloroethane (DDT) has been banned from use in many countries. Other organochlorine pesticides are now largely being replaced with organophosphorus and carbamate pesticides. These newer types are less persistent in the environment, and have not been found to be carcinogenic in experimental settings." The Panel further stated that:

> The use of persistent organic pollutants (organochlorine pesticides, furans, dioxins, and polychlorinated biphenyls) will be banned by 2025 under the United Nations Environment Programme's Stockholm Convention, which entered into force in May 2004.
>
> Many of these contaminants have the potential to accumulate within food systems, and residues of pesticides and herbicides that have been banned from use, or are being phased out, may still be present in foods eaten today. Some contaminants, such as heavy metals and persistent organic compounds, tend to be deposited in fatty tissues and are not easily metabolised or excreted. They accumulate in living creatures, in amounts higher than background levels (for instance, in the soil). Dietary exposure increases with each step up the food chain, as predators consume prey contaminated with these residues.
>
> There are theoretical grounds for concern, which are constantly reviewed by international and national regulatory bodies. However, there is no epidemiological evidence that current exposures are causes of cancers in humans, and so the Panel has made no judgements.

In a section dealing with organic farming (4.9.2) the Panel concluded that:

> Claims that foods produced by organic methods are biologically or nutritionally superior to food produced by intensive methods are not supported by clinical or epidemiological evidence, but some food compositional data indicates higher concentrations of some constituents like vitamin C and dietary fibre. There is evidence that organic products contain fewer residues from chemicals employed in conventional agriculture. However, the subject remains a matter of controversy.

A report published by Gold *et al.* (2002) analyzed studies of the relationship between pesticide residues in food and human cancer development. The authors of this report concluded that there was no conclusive epidemiological evidence that pesticide residues are a significant factor for human cancer. This conclusion supports the findings of the Ad Hoc Panel on Pesticides and Cancer (Ritter, 1997), convened by the National

Cancer Institute of Canada. That Panel also concluded that an increased intake of pesticide residues associated with increased intake of fruits and vegetables did not pose any increased risk of cancer. This agrees with the findings of a University of California (Berkeley) critical analysis (Gold *et al.*, 2001) that there is substantial evidence that high consumption of produce, which may accordingly increase exposure to pesticide residues, has a protective effect against many forms of cancer. In fact, low fruit and vegetable intake has been linked to chronic disease development (World Health Organization, 2003). Consequently agencies such as the Canadian Cancer Society, the Heart and Stroke Foundation and Health Canada recommend that consumers should eat more of these foods.

An editorial in the *Annals of Oncology* (Boyle *et al.*, 2008, reprinted with permission) outlines how the position of the cancer agencies has changed in the 10 years from the publication of their previous report:

> Ten years ago, steps to prevent cancer were known. Intake of vegetables and fruits was clearly associated with reduced cancer risk, antioxidants were a crucial ingredient in lowering cancer risk and overweight and obesity was the most important risk factor for cancer in nonsmokers. What were clear messages then have been confused by subsequent findings and media reports. It is useful to stop and take stock of the current situation.
>
> The current report presents considerably weaker conclusions on the basis of the dilution of the strength of the evidence linking specific components of diet and nutrition to cancer risk. This is particularly true for the fruits and vegetables for which the strength of evidence for a protective effect of high intakes against several common epithelial cancers was downgraded from "convincing" in the first WCRF report to "probable" in this second report (www.wcrf.org and www.aicr.org, reprinted with permission). When considering the importance and significance of diet and, more generally, nutrition on cancer risk, it is essential to separate "what we know" from "what we think we know" from "what we believe" and to understand "what we do not know". "We think we know" or, more accurately, "we thought we knew" that a high-fat diet and low consumption of fruits, vegetables and fibres were associated with increased risks of common cancers. However, faith in the cancer prevention properties of fruits and vegetables began to crack when all the available evidence was critically reviewed by an International Agency for Research on Cancer (IARC) Working Group. Subsequently, it has crumbled as major analyses of prospective studies have continued to demonstrate consistently a lack of association between intake of fruits and vegetables and risk of several cancers. In view of the fragile grounds on which the conclusions of WCRF report on diet and cancer are based on, the information to the media should have been more cautious. Evidence from animal and physiological experiments,

although with conflicting evidence from observational studies, led to antioxidants being widely considered as one of the key protective elements in the diet, being found commonly in vegetables and fruits, and gave rise to a series of chemopreventive trials. A recent Cochrane Review has thrown light on the use of antioxidant supplementation on mortality and found no evidence to support antioxidant supplementation (e.g. vitamin A, beta-carotene, vitamin C, vitamin E, selenium) for either primary or secondary prevention. Indeed, vitamin A, beta-carotene and vitamin E may increase mortality. After decades of research activity, we still do not know how we need to change what we eat to reduce our cancer risk.

As pointed out by Trewavas (2004), farmers, foresters, pesticide users and manufacturers are by occupation more likely to be exposed to higher pesticide hazards than the general public. Consequently many published studies have investigated cancer rates in these groups using comparisons with matched controls from the public, particularly in age and social status. Of 12 separate investigations on farmers involving in total about 300 000 people, 11 found that farmers had overall cancer rates very substantially lower than the general public. Slightly lower numbers of investigations on pesticide users and foresters revealed a similar (8 out of 11) trend whilst only with pesticide manufacturers were cancer rates similar to those of members of the public. The reasons why farmers are so healthy are not known, but Trewavas (2004) suggested that these data indicate not only no evidence for a negative relationship in farmers between pesticide exposure and cancer but a consistent result for the alternative, that pesticide exposure may protect against cancer. He pointed out that farmers do have higher rates of the less common lymphoma but evidence indicates that lymphoma is unrelated to pesticide exposure. Higher exposure to animal disease viruses or fungal disease may be a more plausible hazard. Another point made by Trewavas (2004) was that the stomach is the tissue most likely to be substantially exposed to ingested pesticides but stomach cancer rates have declined by about 60 percent in the last 50 years in Western countries.

Other research on food and health

As indicated above and in previous chapters, conventional food is just as safe as organic food. The documented evidence indicates that risks from pesticide and chemical residues and from microbes are broadly similar in both types of food. Levels of some chemical and pesticide residues may be higher in conventional foods than in organic foods but the documented levels are so low as to be unimportant in terms of affecting human health. Mycotoxin levels are likely to be higher in organic grains than in conventional grains. Because of strict regulations, milk is of high quality whether from cows treated with bovine growth hormone or not. Only raw milk poses any danger to the consumer, and the risk applies to both organic and conventional milk.

The possible benefits of organic food over conventional produce in terms of human health have not been well researched. Demonstrating that dietary exposure to pesticide residues at levels typically found in and on foods is harmful to the majority of consumers is difficult to prove. For instance, if consumers are being exposed to dangerous levels of pesticides in vegetable produce, how can that be reconciled with the fact that intake of fruit and vegetables is higher now than in the past yet life expectancy is also increasing? Consequently, any conclusion would be largely speculative. Another important fact is that there have been no obvious effects on human life expectancy or public health resulting from the introduction of pesticides about 50 years ago and their widespread usage in agriculture and horticulture since then (Coggon and Inskip, 1994). Life expectancy continues to increase, at least in developed countries.

Several bodies, in addition to the cancer agencies, have commented on the health claims for organic foods. For instance at the Food Safety Authority of Ireland (FSAI) meeting in 2008, Dr Mary Flynn, chief specialist for Public Health Nutrition, said: "There is a widespread belief among advocates of organic food that ecological, low input production systems result in foods of higher nutritional quality. However, although the nutrition and health aspects of organic food have been explored in many studies, there is little evidence that organic foods confer health benefits to humans."

These official statements conflict with the views held by many consumers who feel that organic food is more "healthful" in that it contains fewer residues of cancer-causing chemicals. For instance, as pointed out earlier in this book, buyers of conventional food estimated the annual fatality rate due to pesticide residues in conventionally grown food in the USA to be about 50 per million population. Buyers of organic food put the figure at 200 per million. As the researchers pointed out, these estimates are similar in magnitude to the annual mortality due to motor vehicle accidents in the United States (120–160 people per million people or around 100 per day). Do consumers really feel that pesticide residues in food are as dangerous as car crashes in causing death in humans? Apparently so!

It is clear that a main concern often voiced by consumers is the possible presence of chemical and pesticide residues in food. For instance, Mondelaers *et al.* (2009) found that health-related traits were more important than environmental traits in shaping consumer preference for organic vegetables. The presence of an organic label was important in relation to buying intensity.

One of the few studies that investigated the effect of organically grown food on human health was by Schuphan (1974), whose studies extended from 1960 to 1972. He found that organic manuring alone of vegetables was insufficient to obtain a normal yield and optimum nutritional value. He showed an increased content of nutrients grown with manure + fertilizer over that with fertilizer alone, but his figures indicate that the increased content could be explained by a higher dry matter (lower moisture) content. He reported that in experiments with infants who were fed vegetables grown with different fertilizer treatments a combination of stable manure + fertilizer gave the best nutritional results. How he arrived at this conclusion is not clear. This author made one sage comment in relation to how food should be produced: "Emotions prevail instead of knowledge derived from thorough scientific investigations." A person has merely to conduct a search for "organic vs. conventional food" on the internet to prove the wisdom of this statement.

Another area of research on the topic of a possible relationship between organic food and human health is the incidence of allergies in infants. Kummeling *et al.* (2008) investigated whether organic food consumption by infants was associated with developing allergies in the first 2 years of life. Diet was defined as conventional (less than 50 percent organic), moderately organic (50–90 percent organic) and strictly organic (more than 90 percent organic). The records showed that of the total of 815 infants, 10 percent had consumed a moderately organic diet and 6 percent a strictly organic diet. Blood samples taken at 2 years of age were analyzed for IgE (the immunoglobulin that triggers the most powerful allergic reactions). Eczema was present in 32 percent of infants, recurrent wheezing in 11 percent and prolonged wheezing in 5 percent. At 2 years of age, 27 percent of children were found to have become sensitized against at least one allergen. Consumption of organic dairy products was associated with lower eczema risk, but there was no association of organic meat, fruit, vegetables or eggs, or of the proportion of organic products within the total diet, with the development of eczema, wheezing or other signs of allergy. These results do not indicate a strong relationship between consumption of organic foods and a lower incidence of allergies.

Although it is assumed that higher levels of phenolic compounds and antioxidants in plant tissues are beneficial in the human diet, there is a lack of definitive information on this topic. One of the few research groups working in this area (Olsson *et al.*, 2006) studied the effects of extracts from strawberries of cv. Honeoye, grown both organically and conventionally, on cancer cell proliferation *in vitro*. The contents of several antioxidants in the strawberry extracts were analyzed and their inhibition effect on cancer cell proliferation was tested. Differences were found in the content of the antioxidants between the two cultivation methods. The content of ascorbate was 36 percent higher and the ratio of ascorbate to dehydroascorbate was eight times higher in the organic strawberries than in the conventional strawberries, whereas the contents of ellagic acid, total anthocyanidins and total phenolics were lower in the organic strawberries. No significant differences were found in the content of hydroxycinnamic acid and flavonols. The strawberry extracts inhibited cell proliferation in colon cancer cells HT29 and breast cancer cells MCF-7 in a concentration-dependent manner. Extracts from organic strawberries inhibited cell proliferation to a higher extent than from conventional strawberries at the two highest concentrations, suggesting that the content of secondary metabolites with anticancer properties were higher in the organic strawberries. It was hypothesized that the higher content of ascorbate found in the organic strawberries might have affected the higher inhibition of cancer cell proliferation by the extracts from the organic strawberries. Ascorbate is thought to act synergistically with other substances in the extracts.

Briviba *et al.* (2007) studied the effect of consumption of organically and conventionally produced apples on antioxidant activity and DNA damage in human peripheral blood lymphocytes. Six healthy volunteers consumed either organically or conventionally grown apples (Golden Delicious, 1000 g) from two neighboring commercial farms in a double-blinded, randomized, cross-over study. The average content of total identified and quantified polyphenols in the organically and conventionally produced apples was 308 and 321 µg/g fresh weight, respectively. No statistically significant differences in the total content of phenolic compounds or in either of the polyphenolic classes were found between the production methods. Consumption of neither organically nor conventionally grown apples caused any changes in antioxidant capacity of low-density lipoproteins (lag time test), endogenous DNA strand breaks, Fpg protein-sensitive sites, or capacity to protect DNA against damage caused by hydrogen peroxide. However, a statistically significant decrease in the levels of endonuclease III sensitive sites and an increased capacity to protect DNA against damage induced by iron chloride were determined 24 hours after consumption of both types of apples, indicating a similar antigenotoxic potential in organically and conventionally grown apples.

While this area of research is of great potential value in helping to improve the health and well-being of the human consumer, there are at present insufficient findings to allow a clear conclusion to be drawn.

Health of farmers and farm workers

Given the inadequacies in the current data, researchers have attempted to find other ways of assessing human health in relation to organic food. One way of examining the possibility that organic food is more "healthful" is by assessing the health of organic farmers. Logically they should be healthier and live longer than other farmers. However, it does not appear that such data are available at present in the vital statistics tables published in any country. Life expectancy in the general public continues to increase and farmers as a group appear to be long-living, as noted above.

Researchers have therefore investigated other measures of health in organic farmers. One such area of research is semen quality, since some chemicals in the environment are known to interfere with hormones involved in reproduction. Abell *et al.* (1994) found that organic farmers had a higher sperm density than three groups of blue-collar workers.

Larsen *et al.* (1998) conducted a study on the time to pregnancy and exposure to pesticides in Danish farmers. The logic for the work is that toxicological effects on spermatogenesis have been described in humans and animals after exposure to several pesticides. These researchers pointed out that circumstantial evidence suggests that organic farmers may have a higher sperm count than other men, but that comprehensive epidemiological studies of male fertility among farmers have never been carried out. A substantial increase of sperm count is expected to translate into a shorter time to pregnancy, i.e. the number of menstrual cycles or months it takes a woman to get pregnant after discontinuation of birth control. The aim of this study was to examine time to pregnancy among farmers who used pesticides (traditional farmers) and farmers who did not (organic farmers). Data were collected on a total of 904 men. No overall effect of pesticides on male fertility was found; also the researchers found no evidence of higher male fertility in organic farmers.

The same research group (Larsen *et al.*, 1999) carried out a subsequent study to confirm or refute the hypothesis that organic farmers have higher sperm concentrations than traditional farmers. Traditional and organic farmers were selected randomly from central registers, and 171 traditional farmers and 85 organic farmers delivered one semen sample before the start of the spraying season. The traditional farmers had a significantly lower proportion of normal spermatozoa, but this result was not confirmed in a second sample. The researchers concluded that, despite slight differences in concentrations of reproductive hormones, no significant differences in conventional measures of semen quality were found between organic and traditional farmers.

Another study by the same research group (Juhler *et al.*, 1999) investigated whether semen quality was affected by organic food consumption or exposure to pesticide residues in the diet. Data obtained from farmers were divided into three groups according to the amount of organic food consumed and estimated dietary pesticide intakes (of 40 compounds). Although the pesticide intake was found to be lower in the "high organic food consumption" group, the pesticide intake of all groups was estimated to be very low. The group of men who consumed no organic food was found to have a significantly lower proportion of morphologically normal semen but, for the other 14 semen parameters measured, no significant differences were found.

Thonneau *et al.* (1999) conducted a study to determine whether there was a relationship between male exposure to pesticides and the amount of time needed by farmers and agricultural workers in France and Denmark to conceive. The authors used retrospective studies to compare the time to pregnancy of couples in which the man was exposed to pesticides during the year before the birth of their youngest child with that of couples in which the man was not exposed. In 1995 and 1996 the authors studied 362 French rural workers (142 exposed to pesticides and 220 not exposed), 449 Danish farmers (326 conventional farmers exposed to pesticides and 123 non-exposed organic farmers), and 121 Danish greenhouse workers exposed to pesticides. This study found no relationship between time to conception and male exposure to pesticides.

Abell and Bonde (2000) conducted a study on semen quality and sexual hormones in greenhouse workers. The objective was to test whether testicular function was affected by exposure to pesticides. Semen was examined for 122 of 199 eligible men (61 percent) from 30 ornamental flower greenhouses. Sperm concentration, morphology and viability were measured according to World Health Organization guidelines, and the curvilinear sperm velocity was determined by a computer-assisted analysis of video recordings. Three groups were formed according to expert judgment of current exposure to pesticides from cultures, pesticide formulations, and the transfer of pesticide residues from leaves to hands, and also ranked according to years of work in a greenhouse. The risk estimates were adjusted for the effects of sexual abstinence and other potentially confounding factors. The results showed that, based on current exposure, the median values of sperm concentration and the proportion of normal spermatozoa were 60 and 14 percent lower, respectively, in the high-level exposure group ($n = 13$) than in the low-level group ($n = 44$), and the values of the intermediate group fell in between. The adjusted differences between the high-level and low-level exposure groups were statistically significant, while no differences were observed for the viability and velocity of sperm and sexual hormones. The median sperm concentration was 40 percent lower for the men with more than 10 years' experience in a greenhouse than for those with less than 5 years' experience. The age-adjusted testosterone/sex-hormone-binding globulin ratio declined 1.9 percent per year of work. It was concluded by the researchers that male fertility might be at risk from exposure to pesticides in the manual handling of cultures in greenhouses.

In conclusion, these studies do not provide strong evidence of any effect of organic food consumption or pesticide exposure on semen quality and hormone status (except in greenhouse environments), although the researchers concluded that sperm concentration could be investigated further. Perhaps the obvious conclusion is that organic farmers may not be healthier but they are sexier!

Other approaches

Another approach to the question of organic food and health is to conduct experiments on test animals. Baranska *et al*. (2008) conducted a study with rats to examine whether an organic diet affected the immune response. The diets contained lactoalbumin, casein, rapeseed oil, minerals and vitamins, and were analyzed for total flavonoids, polyphenols, β-carotene and lutein, pesticides, and antioxidant activity. Adult male and female Wistar rats were kept for 3 weeks under controlled conditions with free access to water and feed. The test diets were:

- without pesticides and mineral fertilizers;
- without pesticides;
- without mineral fertilizer;
- with pesticides and mineral fertilizers; and control (conventional) feed.

The animals were paired and bred. Twelve progeny (young males) from each experimental group were then fed each of the diets for 3 and 12 weeks. Higher lymphocyte (white blood cell) proliferation was found in rats on both diets based on feeds that had been grown without the use of mineral fertilizers. The responses correlated with the content of flavonoids and polyphenols in the diets. Proliferation was lower in rats fed diets containing organic feeds. The researchers speculated that the alteration in lymphocyte proliferation could be the result of an immuno-stimulatory effect of flavonoids and polyphenols in the diet or from the immunosuppressive influence of some dietary components in feedstuffs grown with mineral fertilizers.

Some contrary findings on the value of plant-based antioxidants in the diet in relation to human health were provided by Hsieh *et al.* (2010), adding to the argument above that the role of antioxidants in the diet is controversial. These researchers studied quercetin, especially abundant in onions and black tea, and ferulic acid, found in corn, tomatoes, and rice bran. They found that diabetic laboratory rats fed either quercetin or ferulic acid developed advanced forms of kidney cancer, and concluded that the two antioxidants appear to aggravate or possibly cause kidney cancer. These researchers concluded that health agencies such as the US Food and Drug Administration should re-evaluate the safety of plant-based antioxidants.

Conclusions

The conclusions of Williams (2002) are still pertinent to the whole issue of organic food and the health of consumers:

> There are virtually no studies of any size that have evaluated effects of organic v. conventionally grown foods on human health.
> The published findings indicate that it is highly unlikely there is any difference in occurrence of chemical residues of drugs, vaccines, pesticides, antibiotics and/or growth promotants in conventional and organic meat (including fish), milk or eggs.

Observational studies that have been conducted to compare the health profiles of organic and conventional farmers are of questionable value because of the possibility of confounding by other lifestyle factors.

The quality and quantity of the science applied in this area to date is inadequate. Conclusions cannot be drawn regarding potentially beneficial or adverse nutritional consequences, to the consumer, of increased consumption of organic foods.

It is apparent that there needs to be better documentation of chemical residue levels in conventional and organic foods before it can be argued conclusively that organic foods are better in this respect. This was clear from a 2005 joint Israeli–Belgian review of the scientific evidence entitled "Need for research to support consumer confidence in the growing organic food market" (Siderer *et al.*, 2005). After reviewing the relevant scientific reports on a comprehensive range of foods (cereals and cereal products, potatoes, vegetables and vegetable products, fruit and fruit products, wine, beer, bread, milk and dairy products, meat and egg products, eggs and honey), the authors concluded that there has been very little published information on residue levels, insufficient to allow clear conclusions to be drawn.

An important finding is that farmers have overall cancer rates very substantially lower than that of the general public. This is contrary to what would be expected if there was a strong correlation between pesticide usage and incidence of cancer.

To date the studies do not provide strong evidence of any effect of organic food consumption or pesticide exposure on semen quality and hormone status (except in greenhouse environments), although the researchers concluded that the issue of sperm concentration could be investigated further.

References

Abell, A. and Bonde, J.P. (2000). Semen quality and sexual hormones in greenhouse workers. *Scandinavian Journal of Work and Environmental Health* **26**, 492–500.

Abell, A., Ernst, E. and Bonde, J.P. (1994). High sperm density among members of organic farmers' association [letter]. *Lancet* **343**, 1498.

Baranska, A.M., Rembialkowska, E., Hallmann, E., *et al.* (2008). The effects of organic v. conventional diets on immune variables in rats. *Proceedings of the Nutrition Society* **67**, E28.

Boyle, P., Boffetta, P. and Autier, P. (2008). Diet, nutrition and cancer: public, media and scientific confusion. *Annals of Oncology* **19**, 1665–1667.

Briviba, K., Stracke, B.A., Rufer, C.E., *et al.* (2007). Effect of consumption of organically and conventionally produced apples on antioxidant activity and

DNA damage in humans. *Journal of Agricultural and Food Chemistry* **55**, 7716–7721.

Coggon, D. and Inskip, H. (1994). Is there an epidemic of cancer? *British Medical Journal* **308**, 705–708.

ECDC-EFSA (2011a). Bean sprouts confirmed as deadly *E. coli* source. Joint Technical Report, European Centre for Disease Control (ECDC) and European Food Safety Authority (EFSA), Berlin, Germany, June 10, 2011. http://www.foodproductdesign.com/news/2011/06/bean-sprouts-confirmed-as-deadly-e-coli-source.aspx, accessed June 10, 2011.

ECDC-EFSA (2011b). *Joint Rapid Risk Assessment: Cluster of Haemolytic Uremic Syndrome (HUS) due to Escherichia coli (E. coli) O104:H4 in Bordeaux, France* . Stockholm, Sweden, European Centre for Disease Prevention and Control.

EU (2011). *E. coli* outbreak in Germany. European Commission Report IP/11/653, Brussels, May 27, 2011.

Gold, L.S., Slone, T.H., Ames, B.N. and Manley, N.B. (2001). Pesticide residues in food and cancer risk: A critical analysis. In: *Handbook of Pesticide Toxicology*, Second Edition (R. Krieger, ed.). Academic Press, San Diego, CA, pp. 799–843. http://potency.berkeley.edu/text/handbook.pesticide.toxicology.pdf, accessed September 22, 2004.

Gold, L.S., Slone, T.H., Manley, N.B. and Ames, B. N (2002). *Misconceptions about the Causes of Cancer*. The Fraser Institute Centre for Studies in Risk, Regulation and Environment, Vancouver, BC, Canada. http://potency.berkeley.edu/text/Gold_Misconceptions.pdf, accessed September 22, 2004.

Hord, N.G., Tang, Y. and Bryan, N.S. (2009). Food sources of nitrates and nitrites: the physiologic context for potential health benefits. *American Journal of Clinical Nutrition* **90**, 1–10.

Hsieh, C.-L., Peng, C-C., Cheng, Y-M., *et al.* (2010). Quercetin and ferulic acid aggravate renal carcinoma in long-term diabetic victims. *Journal of Agricultural and Food Chemistry* **58**, 9273–9280.

Juhler, R.K., Larsen, S.B. and Meyer, O. (1999). Human semen quality in relation to dietary pesticide exposure and organic diet. *Archives of Environmental Contamination and Toxicology* **37**, 415–423.

Katan, M.B. (2009). Nitrate in foods: harmful or healthy? *American Journal of Clinical Nutrition* **90**, 11–12.

Kummeling, I., Thijs, C., Huber, M., *et al.* (2008). Consumption of organic foods and risk of atopic disease during the first 2 years of life in the Netherlands. *British Journal of Nutrition* **99**, 598–605.

Larsen, S.B., Joffe, M. and Bonde, J.P. (1998). Time to pregnancy and exposure to pesticides in Danish farmers. *Occupational and Environmental Medicine* **55**, 278–283.

Larsen, S.B., Spano, M., Giwercman, A. and Bonde, J.P. (1999). Semen quality and sex hormones among organic and traditional Danish farmers. *Occupational and Environmental Medicine* **56**, 139–144.

Mondelaers, K., Verbeke, W. and Van Huylenbroeck, G. (2009). Importance of health and environment as quality traits in the buying decision of organic products. *British Food Journal* **111**, 1120–1139.

Olsson, M.E., Andersson, C.S., Oredsson, S., *et al.* (2006). Antioxidant levels and inhibition of cancer cell proliferation in vitro by extracts from organically and conventionally cultivated strawberries. *Journal of Agricultural and Food Chemistry* **54**, 1248–1255.

Pussemier, L., Larondelle, Y., Carlos, V.P.C. and Huyghebaert, A. (2006). Chemical safety of conventionally and organically produced foodstuffs: a tentative comparison under Belgian conditions. *Food Control* **17**, 14–21.

Ritter, L. (1997). Report of a panel on the relationship between public exposure to pesticides and cancer. *Cancer* **80**, 2019–2033.

Schuphan, W. (1974). Nutritional value of crops as influenced by organic and inorganic fertilizer treatments. Results of 12 years' experiments with vegetables (1960–1972). *Plant Foods for Human Nutrition* **23**, 333–358.

Siderer, Y., Maquet, A. and Anklam, E. (2005). Need for research to support consumer confidence in the growing organic food market. *Trends in Food Science and Technology* **16**, 332–343.

Sobko, T., Marcus, C., Govoni, M. and Kamiya, S. (2010). Dietary nitrate in Japanese traditional foods lowers diastolic blood pressure in healthy volunteers. *Nitric Oxide* **22**, 136–140.

Thonneau, P., Abell, A., Larsen, S.B., *et al.* (1999). Effects of pesticide exposure on time to pregnancy: results of a multicenter study in France and Denmark. *American Journal of Epidemiology* **150**, 157–163.

Trewavas, A. (2004). Review. A critical assessment of organic farming-and-food assertions with particular respect to the UK and the potential environmental benefits of no-till agriculture. *Crop Protection* **23**, 757–781.

Williams, C.M. (2002). Nutritional quality of organic food: shades of grey or shades of green? *Proceedings of the Nutrition Society* **61**, 19–24.

World Cancer Research Fund/American Institute for Cancer Research (2007). *Food, Nutrition, Physical Activity, and the Prevention of Cancer: a Global Perspective*. American Institute for Cancer Research, Washington, DC.

World Health Organization (2003). *Social Determinants of Health: the Solid Facts*, Second Edition (R. Wilkinson and M. Marmot, eds). World Health Organization, Copenhagen.

10 Is Organic Food More Nutritious and "Tasty"?

A main reason why consumers buy organic food – apart from the safety aspect – is that it is perceived to be of higher quality. This aspect is an interesting one because consumers often equate freshness with quality. Organic food produced and purchased locally is likely to be fresher than food shipped over a distance, perhaps from another country. This is a main reason for the popularity of farmers' markets.

The preference for fresh food is stronger in Europe than in North America and this may be one reason why the organic industry is larger there. French consumers, for instance, prefer to buy bread every day from the bakery. How many consumers in North America do that?

Another main reason is that, in addition to being fresher, organic food is perceived to be more nutritious. This is a controversial area, with an inadequate amount of scientific data to prove or disprove the point.

Reviews

One of the first comprehensive reviews of reports on the nutritional value of organic foods was by Woese et al. (1997). These authors reviewed data published from 1926, comprising results from more than 150 investigations which compared the quality of conventional and organic foods. The foods included cereals, potatoes, vegetables, fruits, wine, beer, bread, cakes and pastries, milk, meat, eggs and honey, as well as products made from them. Nutritional studies involving animals and humans were included. Overall conclusions were difficult to reach because of the wide variation in the various factors involved and in sampling methods. However some differences were identified. Vegetables that were cultivated or grown with the use of fertilizer had a much higher nitrate content than vegetables grown and fertilized organically. The effect was most noticeable in leaf, root and tuber vegetables such as spinach, lettuce and beets. These vegetables are known as nitrophilic because they take up more nitrogen than can be converted into protein. A trend for this effect was also seen in potatoes.

Organic Production and Food Quality, First Edition. Robert Blair.
© 2012 John Wiley & Sons, Ltd. Published 2012 by John Wiley & Sons, Ltd.

Whether or not dietary nitrate has toxic or beneficial properties for human health is controversial (as discussed in the previous chapter). A higher dry matter concentration (less water) was found in organic vegetables than in conventional products, particularly leaf vegetables. Cereals showed differences in processing properties. Conventionally grown wheat had a higher protein content and a superior protein quality than organic wheat, which made it more suitable for baking. No other differences in nutritional value were observed, or the results were contradictory and did not allow a clear distinction to be made. The same applied to sensory tests.

Other scientists have attempted to answer the question about nutritional content by assessing the published findings. Worthington (2001) surveyed the available literature comparing nutrient content of organic and conventional crops using statistical methods to identify significant differences and trends in the data. She concluded that organic crops contained significantly more vitamin C, iron, magnesium, and phosphorus and significantly less nitrate and heavy metals than conventional crops. There was a trend for a lower content of protein, but it was of a higher quality. Her overall conclusion was that there appeared to be genuine differences in the nutrient content of organic and conventional crops in favor of organic crops.

A more extensive review was conducted by researchers in New Zealand (Bourn and Prescott, 2002). They attempted to resolve this issue by assessing the scientific findings relating to organically and conventionally grown foods in three key areas, namely nutritional value, sensory quality, and food safety. Their assessment concluded that there were few well controlled studies that allowed a valid comparison. With the possible exception of nitrate content, there was no strong evidence that organic and conventional foods differed in nutrient content. While there were reports indicating that organic and conventional fruits and vegetables differed in a variety of sensory qualities, the findings were inconsistent.

A similar conclusion was reached by a researcher at the Hugh Sinclair Unit of Human Nutrition at the University of Reading, UK (Williams, 2002). She pointed out that there appears to be a widespread perception among consumers that organic production methods result in foods of higher nutritional quality, but she found a lack of evidence in the scientific literature to support or refute this perception. Very few differences in food composition had been reported, although there were reasonably consistent findings for higher nitrate and lower vitamin C contents of conventionally produced vegetables, particularly leaf vegetables.

Other scientists and agencies have reached similar conclusions. The UK, French and Swedish government food agencies all concluded that organic food is not significantly different in terms of food safety and nutrition from food produced conventionally. Alex Avery, a scientist and research director of the Center for Global Food Issues at the Hudson Institute in the United States, published a book, *The Truth About Organic Foods*, in which he refuted claims that organic food is healthier than conventional food (Avery, 2006). In support of his conclusion he quoted William Lockeretz (an organic advocate) who, in 1997, told an organic food conference:

> I wish I could tell you that there is a clear, consistent nutritional difference between organic and conventional foods. Even better, I wish I could tell you that the difference is in favour of organic. Unfortunately, though, from my reading of the scientific literature, I do not believe that such a claim can be responsibly made.

On the other hand organic producer organizations such as the Soil Association promote the view that organic food is more healthful. For instance, the Soil Association website states:

> Many people believe organic food is better for you because not only is it produced avoiding pesticides and contains far fewer additives, but also because there is increasing evidence it contains more beneficial nutrients. A rapidly growing body of research shows organic food contains higher levels of vitamin C and essential minerals such as calcium, magnesium, iron and chromium as well as cancer-fighting antioxidants and Omega 3. For example, studies show organic milk is on average 68 percent higher in Omega 3 essential fatty acids. It's thought this is because of the high levels of natural red clover fed to cows on organic dairy farms.
> (*Source:* Research by Universities of Liverpool and Glasgow from 2002 to 2005 published in the *Journal of Dairy Science* in 2006.)

This is an important issue for the consumer. *The Toronto Globe and Mail* (a Canadian newspaper) and CTV (a Canadian television station) therefore commissioned a research study in 2002 at the University of Guelph (Ontario) to try to find out whether organic foods are superior in nutrient content. Researchers at the University of Guelph food laboratory bought 20 types of organic produce from a Toronto health-food outlet, analyzed them, and compared the results with tables produced by Health Canada on the nutrient content of foods.

There was no appreciable difference between the nutritional content of organic fruits and vegetables and conventionally grown ones, according to Andre Picard in reporting the results in the *Globe and Mail*. Of 135 tests conducted, conventional produce came out ahead 66 times and organic produce 49 times. Organic potatoes, for example, had three times the iron of conventional potatoes, but only half the calcium. Organic red pepper had 42 percent more vitamin C than conventional red pepper, but 45 percent less vitamin A.

The report went on to state that the results were no surprise to Phil Warman, an agronomist and professor of agricultural sciences at Nova

Scotia Agricultural College. He had been growing organic and conventional crops on a test farm for 12 years, and reported that the results were always the same.

"In terms of nutritional content, there is virtually no difference," Dr Warman said. "I know this is disappointing for organic growers to hear this and probably for the consumer who has been led to believe the food is nutritionally superior."

The findings were not received well by supporters of the organic industry, according to the newspaper article. Donna Herringer, president of the Canadian Health Food Association and a long-time consumer of organic foods, was reported as being indeed disappointed to hear the results, but not the least bit swayed in her belief that organic is superior.

"In the reports I'm reading, organic is always better. Consumers are not buying organic food merely for what is in them (nutrients)", she said, "but for what isn't (pesticides and herbicides). Organics also taste better and are produced by small growers, not multinational corporations," Ms Herringer said. "The most important thing for me is that organics are real. A little bit less vitamin C in one fruit isn't going to make a difference because, over all, the food I eat is way healthier."

In the article, Leonard Piché, an associate professor in the nutrition program at Brescia College in London, Ontario, was quoted as dismissing that view as based on wishful thinking, not science. He said the evidence is that there is no difference between organic and conventional, and that is good news for consumers. "Canadians should feel confident that regardless of whether they purchase organically grown or conventionally grown produce, they can expect to obtain similar nutrient value," he said.

The article also made the point that most nutritionists and public-health experts stress that Canadians need to eat far more fruits and vegetables, irrespective of whether they are organic or conventional.

More recent reports indicate that the controversy continues over the claim that organic food has an improved quality over conventional food. Unfortunately, two such reports, which reviewed the scientific evidence, were published on the internet and did not allow the methodology and the conclusions to be subjected to peer review by scientists prior to publication. One of these was a report entitled "New Evidence Confirms the Nutritional Superiority of Plant-Based Organic Foods" which was published by The Organic Center (Benbrook *et al.*, 2008). These authors identified all peer-reviewed studies published in the scientific literature since 1980 that compared the nutrient levels in plant-based organic and conventional foods and screened them in two ways for scientific validity. They assessed how the studies defined and selected organic and conventional crops for nutrient-level comparisons.

From 97 published studies, they identified 236 scientifically valid "matched pairs" of measurements that included an organic and a conventional sample of a given food. They took into account, for example, the experimental design of each study, the need for the same cultivars to be planted in both the organic and conventional fields, the degree of differences in soil types and topography, the focus of the study and where it was carried out, the definition of organic farming, and years for which the organic field in a matched pair had been managed organically.

For each crop addressed in a given study, they determined whether the study was "high quality", "acceptable" or "invalid" based on explicit inclusion and exclusion criteria and a rating system. They also screened the 94 valid study-crop combinations for the accuracy and reliability of the analytical methods used to measure nutrient levels. This allowed the researchers to compare the levels of 11 nutrients in organic and conventional foods. The nutrients included: four types of antioxidants (total phenolics, total antioxidant capacity, quercetin and kaempferol); three precursors of key vitamins (vitamins A, C and E); two minerals (potassium and phosphorus); nitrates; and total protein.

The researchers found that within the 236 valid matched pairs, the organic foods were nutritionally superior in 145 (61 percent), while the conventional foods were more nutrient dense in 87 (37 percent). There were no differences in 2 percent of the matched pairs. The organic samples contained higher concentrations of polyphenols and antioxidants in about three-quarters of the 59 matched pairs representing those four phytonutrients. Increasing intakes of these nutrients is a vital goal to improve public health since daily intakes of antioxidants and polyphenols are less than one-half of recommended levels. Matched pairs involving comparisons of potassium, phosphorus, and total protein levels accounted for over three-quarters of the 87 cases in which the conventional samples were nutritionally superior. While a positive finding, these three nutrients are clearly of lesser importance than the other eight nutrients because, in general, the former are adequately supplied in the average American diet. The magnitude of the differences in nutrient levels strongly favored the organic samples. One-quarter of the matched pairs in which the organic food contained higher levels of nutrients exceeded the level in the conventional sample by 31 percent or more. Only 6 percent of the matched pairs in which the conventional sample was more nutrient dense exceeded the levels in the organic samples by 31 percent or more. Across all 236 matched pairs and 11 nutrients, the nutritional premium of the organic food averaged an impressive 25 percent.

That review and its conclusions were criticized in a report from the American Council on Science and Health, New York, authored by Joseph D. Rosen (2008), Emeritus Professor of Food Toxicology at Rutgers University's School of Environmental and Biological Sciences. This author explained some of the background to the selection of nutrients listed in the Benbrook *et al.* (2008) report.

> Free radicals are reactive chemical species produced during normal human metabolism. It is widely believed that they can initiate cancer by reacting with and damaging DNA. Free radicals can also

contribute to coronary heart disease by oxidizing "bad cholesterol" to form arterial plaque. Antioxidants are capable of destroying free radicals, and that's why antioxidants are of so much interest. Health experts advise us to eat five to nine servings of fruits and vegetables every day because these foods are rich in vitamin C, vitamin E, beta-carotene, lycopene and chemicals known as polyphenols. All of these chemicals are powerful antioxidants. Two other powerful antioxidants, butylated hydroxytoluene and butylated hydroxyanisole, cannot be used in organic food because they are synthetic chemicals.

Some polyphenols are produced by plants in response to attacks by insects, fungi and weeds (among other stresses), and it is believed by some scientists that organic crops produce more of these chemicals than conventional crops because the latter are already protected against some of these stresses by synthetic pesticides. Flavonoids are a subclass of polyphenols and a great number of these are found in food. Two flavonoids highlighted in The Organic Center Report are quercetin and kaempferol. There are a large number of polyphenols in food and it is very difficult to measure all of them individually. Therefore total polyphenols (or total phenolics, as they are sometimes called) are measured instead of, or in addition to, individual polyphenols.

A second theory is that slower release of nitrogen, as occurs when manure is substituted for synthetic fertilizer, results in higher polyphenol concentrations in food crops (reprinted with permission of the author).

Rosen criticized the methodology in The Organic Center report, pointing out that one important variable not taken into account with the crop data was the change in nutrient content between growing seasons. Another was that the time from harvesting to market may have been different for the organic and conventional foods. One study on the nutrient content of kiwi fruits included the skin, which most consumers do not eat, as well as the pulp. He drew attention to the claim that organically grown vegetables had more quercetin than conventional varieties and pointed out that the organic vegetables studied had been sprayed with an organic pesticide that would have increased the production of quercetin in these plants. In addition he criticized the way in which the results were interpreted by Benbrook *et al.*, for instance the relevance of some of the statistically significant findings.

Statistical significance is certainly important, but sometimes statistically significant results are not meaningful. While it may be true that milk from cows raised on grass contained 46 percent more vitamin E and 50 percent more beta-carotene than grain-fed cows, it is also true that neither of these nutrients is found to any great extent in milk. An increase of 46 percent of vitamin E is the same as approximately 876 micrograms per quart of milk or less than 6 percent of the

RDA (recommended daily allowance) for this vitamin. Increasing beta-carotene content of milk by 50 percent gets you approximately an extra 112 micrograms per quart, a lot less than the approximately 17 000 micrograms found in one medium size baked sweet potato.

The overall result of his treatment of the data in the Benbrook *et al.* report is shown in Table 10.1.

Whereas Benbrook *et al.* concluded that the organic foods in question had 25 percent more nutrients than conventional foods, Rosen calculated that the nutrient content was virtually the same.

It is understandable, then, that observers of the organic scene are confused over this issue when experts cannot agree on an interpretation of the same data. Commentators on this issue, such as television and newspaper reporters, have to decide on which opinion to accept when filing a story. Much of this confusion could be avoided if, as pointed out here, reports on important topics such as this were critiqued by peer review prior to publication. This would allow the disagreements to be addressed

Table 10.1 Differences in the nutrient content in organic and conventional foods: a comparison between results published by The Organic Center and the Rosen critique

Nutrient	Average ratio of organic to conventional values before corrections (Benbrook *et al.*, 2008)	Average ratio of organic to conventional values after corrections (Rosen, 2008)
Antioxidants		
Total phenolics	1.10	1.03
Antioxidant capacity	1.24	1.06
Quercetin	2.40	1.03
Kaempferol	1.05	1.02
Vitamins		
Vitamin C/ascorbate	1.10	1.10
Beta-carotene	0.92	0.95
Vitamin E	1.15	1.01
Minerals		
Phosphorus	1.07	1.07
Potassium	1.00	1.00
Other nutrients		
Nitrate	1.80	0.56
Protein	0.90	0.90
Average ratio overall	1.25	0.98

(See Rosen (2008) and Benbrook *et al.* (2008). Reprinted with permission of The Organic Center.)

prior to publication. Also, the reports should be published in journals which are recognized for their high standards in accepting papers for publication.

In addition, some simple steps could be taken by authors to make their findings more easily understood by experts and the non-expert. For instance, the nutrient values in fruits and vegetables could be expressed on a similar moisture basis. At present some authors do this on a fresh weight basis, disregarding how much moisture (water) content is present. A piece of fruit may appear to have a lower level of a nutrient, perhaps because it has a higher water content, but when the water content is standardized the nutrient content may be the same. The consumer eating the two pieces of fruit would consume a similar level of nutrient in each case. It has been shown that the water concentration is lower in organic food, especially vegetables and in particular leaf vegetables, in comparison with conventionally grown. High usage of fertilizer stimulates the yield of crops partly by increasing the moisture content.

Another piece of information that would help to explain a possibly higher level of nutrients in organic crops is the total harvested yield. If a crop has a higher nutrient content, can this be explained by the plants producing a similar amount of nutrients to conventional plants, but putting them into fewer fruits or grains? Measurement of the total harvest would allow this possible explanation to be explored. If organic produce can be shown conclusively to have a higher nutrient content than conventional produce, the enhancement has to have an explanation.

One research team that did report total yield as well as nutritional content was that of Warman (the same Dr Warman as cited earlier in this section) and Havard (1997), at the Nova Scotia Agricultural College in Canada. The test involved two treatments, organic and conventional, for carrots (*Daucus carota* L. cv. Cellobunch) and cabbages (*Brassica oleracea* L. var. *capitata* cv. Lennox), in each of three years. The addition of pesticides, lime and NPK (nitrogen, phosphorus, potash) fertilizer to the conventional plots followed soil test and provincial recommendations. Lime, composted manure and insect control applications to the organic plots were according to the guidelines of the Organic Crop Improvement Association Inc. The compost was analyzed for total nitrogen and applied to provide 170 kg nitrogen per hectare for carrots and 300 kg nitrogen per hectare for cabbages, which assumed 50 percent availability of nitrogen. In addition to marketable yields, carrot leaves and roots and cabbage sections were analyzed for 12 macro- and micronutrients. Vitamins C and E and α- and β-carotene of mature crops were determined. In 2 of the 3 years, vitamin C was also analyzed up to 24 weeks after harvest. Analysis of the 3 years of data showed that the yield and vitamin content of the carrots and cabbages were similar in organic and conventional crops.

The same authors (Warman and Havard, 1998) conducted a similar study on the yield, vitamin and mineral contents of organically and convention-ally grown potatoes and sweet corn (maize) and obtained results very similar to those found with carrots and cabbages. Analysis of 3 years of data showed that the yield and vitamin C content of the potatoes were not affected by treatment. However, the conventionally grown treatment out-produced the organically grown treatment for cv. Pride and Joy corn, but there was no difference between treatments in the yield of cv. Sunnyvee corn or in the vitamin C or E contents of the sweet corn kernels in any year.

In summary, most researchers agree that there are few, if any, differences in the nutritional quality of conventional and organic vegetables. However, the data cited by Benbrook *et al.* and others suggest the possibility that secondary nutrients such as antioxidants may be enhanced in some plants grown organically. This seems to be due to the response of the plants to organic treatments used to protect them from pests. The finding may be of significant importance in relation to human health. Danish researchers (Brandt and Mølgaard, 2001) reported on this topic at the International Conference on the Nutritional Enhancement of Plant Foods, which was held at the John Innes Centre, Norwich, UK, in 2000. These researchers examined the possible differences between organic and conventional plant products and their effects on human health. They concluded that nutritionally important differences relating to contents of minerals, vitamins, proteins and carbohydrates were unlikely, nor were present levels of pesticide residues in conventional products a cause for concern. However, they suggested that the levels of many defense-related secondary metabolites in the diet are lower than optimal for human health. They also concluded that there was ample, but circumstantial, evidence that on average organic vegetables and fruits are likely to contain more of these secondary nutrients than conventional produce, suggesting that organic food from plants may benefit human health more than corresponding conventional plant foods.

An important report was published in 2009 by the UK Food Standards Agency (Dangour *et al.*, 2009a) which confirmed the overall assessment in this chapter on the nutritional equivalency of conventional and organic foods. The report, based on an analysis of the published data by the London School of Hygiene and Tropical Medicine, concluded there was no reason to buy expensive organic food for nutritional reasons. The research assessed the best evidence over the last 50 years. After reviewing 160 studies on the nutritional content of organic foods versus non-organic, it concluded there was no significant difference in vitamins and minerals that are important to human health. A further analysis of more than 50 studies on the health implications found no good evidence that organic food is better for you than non-organic. Dr Alan Dangour, one of the investigators, stated that the report was the most comprehensive review of the health benefits of organic food ever carried out. His report stated:

On the basis of a systematic review of studies of satisfactory quality, there is no evidence of a difference in nutrient quality between organically and conventionally produced foodstuffs. The small differences in nutrient content detected are biologically plausible and mostly relate to differences in production methods.

Gill Fine, FSA Director of Consumer Choice and Dietary Health, is quoted as saying about the report:

> ...there is no need for people to buy highly-priced organic food for the health benefits. The study does not mean that people should not eat organic food. What it shows is there is little, if any, nutritional difference between organic and conventionally produced food and that there is no evidence of additional health benefits from eating organic food.

The report received intense media attention worldwide, which was surprising since the findings were not new. Predictably it was not received well by the organic industry, but there was little in the way of factual rebuttals.

One strong rebuttal was made by Benbrook *et al.* (2009) in the form of a letter to the editor of the *American Journal of Clinical Nutrition*. This journal had published the article by Dangour *et al.* (2009a) which was the basis of the report by the UK Food Standards Agency. The letter (in part) stated:

> Dangour *et al.* considered 162 articles that reported comparisons from field trials, farm surveys, or market basket studies. They excluded 54% of these studies simply because the organic certifying body wasn't stated, thus eliminating many otherwise valid comparison studies. Conversely, they apparently accepted studies with mixed cultivars and breeds because they required only identification of the cultivars or breeds not that they be identical within a study. It is well known that there can be large differences in nutrient concentrations between different cultivars of the same crop ... They also arbitrarily excluded from analysis any nutrient with <10 valid studies.

Dangour responded to the Journal, stating (in part):

> Our review brought together for the first time all peer-reviewed published reports on the nutrient content of organic foods, and our conclusions are based on analysis of data presented in the studies that were categorized as satisfactory quality. Our analysis was based only on data reported in published peer-reviewed articles. We acknowledge that there is no standard way of conducting these analyses, and in response to a request we have reanalyzed the extracted data by using the log of the response ratio as our metric. We now report the results of the reanalysis that replicate our original findings.

The reanalysis did not alter the basic conclusions reached initially by Dangour *et al.* (2009b).

Another response by these researchers was to update the number of reports analyzed and to publish a subsequent paper in 2010 (Dangour *et al.*, 2010). The conclusion of this more extensive review was:

> From a systematic review of the currently available published literature, evidence is lacking for nutrition-related health effects that result from the consumption of organically produced foodstuffs.

It is clear that a major difference in approach is that Dangour *et al.* based their analysis on peer-reviewed papers in scientific journals, whereas Benbrook *et al.* used a wider database for their analysis.

Some members of the public were also upset by the findings of Dangour *et al.*, as evidenced by the flurry of letters received by newspapers. Their disappointment was understandable, given that the findings were contrary to what they had been led to expect. Other responses from the public included an unwillingness to accept findings that organic foods are not markedly superior in nutritional attributes than conventional foods (e.g. Donna Herringer quoted above). Another response from some members of the organic industry was that researchers have not been able to detect and measure vital components that distinguish organic (and particularly bio-dynamic) foods from conventional foods. This is regarded as being a vital force that is perhaps immeasurable by current methods.

Analysis by food group

As outlined in the previous chapters, organic production affects the composition of the various food groups differently. Consequently each food group should be considered separately. Also, less information is available on the nutritional composition of organic milk, meat and eggs than on organic foods derived from plants. This makes generalizations more difficult. The documented findings show the following.

Vegetable produce

Limited published data suggest that there are few differences in nutritional content between organic and conventional produce and that the differences found are not of nutritional importance (Chapter 3). Some studies have shown an increased content of vitamin C, β-carotene and phenolic compounds in organic produce, but a study with humans found that a high intake of these compounds from tomatoes did not lead to an increase in plasma concentration. A study with rats showed no difference in mineral retention when fed organic or conventional produce. Limited testing has been done to compare the taste and flavor of organic vs. conventional produce. Conventionally grown tomatoes were preferred to organic tomatoes but results suggest that, when vegetable produce is compared at the same stage of ripeness, no differences in taste or flavor are recorded.

Fruit

As outlined in Chapter 4, a strong perception held by many consumers is that organic production results in fruit of higher nutritional value. Data on this issue are, however, limited. Also it is not clear from the research studies whether the fruit composition was always compared on a similar moisture basis and at the same stage of ripeness. As with vegetable produce, there is evidence that some organic fruit is drier than conventionally grown fruit. Unless this factor is taken into account, a higher content of a nutrient might be explained by a higher dry-matter (lower moisture) content. A slightly drier fruit may also have a more intense flavor due to the higher concentration of nutrients, and as a result may be preferred by the consumer. There is evidence that some organically grown fruits have a higher resistance to deterioration and better keeping quality, attributed to a lower moisture content.

The available data indicate that organic and conventional fruits do not differ significantly in their content of major nutrients. On the other hand, higher contents of phenolic compounds and antioxidants have been reported in some organic fruit, which might be beneficial in terms of human health. This issue was discussed at the Fourth European Conference on Polyphenols and Health which was held in the UK in 2010. At previous conferences on this topic, many presentations have highlighted the antioxidant activity of polyphenols as a major property relating to health. However, the science of polyphenols has progressed and, according to Hollman (2010), many scientists now agree that the simple view of polyphenols as antioxidants that prevent overall oxidative damage in the body should be revised. There is no doubt that polyphenols are excellent antioxidants *in vitro*, but systemic antioxidant effects in the human body are hard to prove. The question of whether dietary polyphenols exert any positive health effect at all was addressed by a review of more than 130 good-quality human intervention studies with polyphenol-rich foods or extracts. From these studies it was concluded that clinically significant positive effects on flow-mediated dilation (a measure of blood vessel function) and blood pressure and low density lipoprotein cholesterol can be achieved. However, it is not clear yet whether these effects are caused by the polyphenols.

Cereal grains

As outlined in Chapter 5, much less information has been published on the nutritional quality of cereal grains produced organically than on vegetable produce or fruit. This is in spite of the fact that grains are produced for the important animal feeding market as well as for human consumption.

Organic cereals generally have a lower content of protein, but in other nutritional aspects are similar to conventional cereals except for a slightly higher content of phenolic compounds. Animals appear to have the ability to discriminate between organic and conventional sources of cereals. In preference tests, rats selected biscuits made from organic wheat over biscuits made from conventional wheat. Conversely, wild birds preferred conventional wheat seed as a winter feed, a result ascribed to a lower content of protein in the organic wheat.

Meat

As outlined in Chapter 6, organic production affects the different classes of meat animals in different ways. Accordingly it is not possible to generalize on the effect. Also, an effect recorded in all species is a lengthening of the time taken for organically raised animals and poultry to reach market weight, so that the effect of organic production is confounded with an age effect.

Beef animals raised organically grow more slowly and produce leaner carcasses. As a result the meat tends to have less marbling and is less tender. The profile of the fat is altered with organic production (or with grass feeding), with a higher content of PUFAs (in particular CLA) and is regarded as more favorable in terms of human nutrition. Similar findings have been reported with pigs and poultry, the research and consumer findings suggesting that the result is a slightly tougher meat but with an enhanced flavor that is preferred by some consumers (probably an age effect since the organic animals take longer to reach market weight). The main difference between organic (farm-raised) and wild fish is a higher content of fat in the organic fish.

Milk

One aspect of food that is of interest to the consumer is the type of fat present, particularly in foods of animal origin. Health authorities in a number of countries (e.g. WHO, 2003) have recommended a reduction in the intake of saturated fats and an increase in the intake of unsaturated fatty acids, particularly the omega-3 polyunsaturated fatty acids (n-3 PUFA), because they are known to be beneficial to human health. Organic production, or at least forage consumption, is known to result in a higher content of PUFAs in milk, mainly in α-linolenic acid. This finding may be of value in terms of human nutrition, but it has been pointed out that the important PUFAs are DHA (docosahexaenoic acid) and EPA (eicosapentaenoic acid). Another consideration is that any potential benefit will be reduced by the selection of fat-free or skimmed milk rather than whole milk.

Eggs

The authoritative book *Egg Science and Technology* (Stadelman *et al.*, 1995) stated that organic eggs are similar to conventional eggs in nutritive value. It also stated that, if the organic diet is not as well balanced as the conventional commercial diet, the nutritive value of organic eggs tends to be lower. Other textbooks make similar comments.

Eggs are a very nutritious food, containing valuable protein and micronutrients. There is no evidence that the cholesterol content of eggs is lower in organic than in regular eggs, although the fatty acid profile may be more favorable. The documented evidence indicates that organic and conventional eggs are otherwise similar in nutritional characteristics, although organic eggs may be smaller and may be brown-shelled rather than white-shelled.

Taste

Since taste is a sensory attribute and more subjective in its assessment than nutritional composition, which can be measured objectively, the issue of the taste of organic food will be addressed in Chapter 11 which deals with the psychological aspects of organic food choice.

References

Avery, A. (2006). *The Truth About Organic Foods*. Henderson Communications, Chesterfield, MO, USA.

Benbrook, C., Zhao, X., Yanez, J., *et al.* (2008). *New Evidence Confirms the Nutritional Superiority of Plant-Based Organic Foods*. State of Science Review, The Organic Center, Boulder, CO, USA.

Benbrook, C., Davis, D.R. and Andrews, P.K. (2009). Methodologic flaws in selecting studies and comparing nutrient concentrations led Dangour et al to miss the emerging forest amid the trees. *American Journal of Clinical Nutrition* **90**, 1700–1701.

Bourn, D. and Prescott, J. (2002). A comparison of the nutritional value, sensory qualities, and food safety of organically and conventionally produced foods. *Critical Reviews in Food Science and Nutrition* **42**, 1–34.

Brandt, K. and Mølgaard, J. P. (2001). Organic agriculture: does it enhance or reduce the nutritional value of plant foods? *Journal of the Science of Food and Agriculture* **81**, 924–931.

Dangour, A.D., Dodhia, S.K., Hayter, A., *et al.* (2009a). Nutritional quality of organic foods: a systematic review. *American Journal of Clinical Nutrition* **90**, 680–685.

Dangour, A.D., Dangour, E.A., Lock, K. and Uauy, R. (2009b). Reply to DL Gibbon and C Benbrook et al. *American Journal of Clinical Nutrition* **90**, 1701.

Dangour A.D., Lock, K., Hayter, A., *et al.* (2010). Nutrition-related health effects of organic foods: a systematic review. *American Journal of Clinical Nutrition* **92**, 203–210.

Hollman, P.C.H. (2010). The 4th International Conference on Polyphenols and Health. *Nutrition Bulletin* **35**, 183–185.

Rosen, J.D. (2008). *Claims of Organic Food's Nutritional Superiority: A Critical Review*. American Council on Science and Health, New York.

Soil Association website http://www.soilassociation.org/Whyorganic/Health/tabid/59/Default.aspx.

Stadelman, W.J., Newkirk, D. and Newby, L. (1995). *Egg Science and Technology*, 4th edition. CRC Press, Boca Raton, FL.

Warman, P.R. and Havard, K.A. (1997). Yield, vitamin and mineral contents of organically and conventionally grown carrots and cabbage. *Agriculture, Ecosystems and Environment* **61**, 155–162.

Warman, P.R. and Havard, K.A. (1998). Yield, vitamin and mineral contents of organically and conventionally grown potatoes and sweet corn. *Agriculture, Ecosystems and Environment* **68**, 207–216.

WHO (2003). Diet, nutrition and the prevention of chronic diseases. World Health Organization Technical Series Report 916. WHO, Geneva.

Williams, C.M. (2002). Nutritional quality of organic food: shades of grey or shades of green? *Proceedings of the Nutrition Society* **61**, 19–24.

Woese, K., Lange, D., Boess, C. and Bogl, K.W. (1997). A comparison of organically and conventionally grown foods – results of a review of the relevant literature. *Journal of the Science of Food and Agriculture* **74**, 281–293.

Worthington, V. (2001). Nutritional quality of organic versus conventional fruits, vegetables and grains. *Journal of Alternative and Complementary Medicine* **7**, 161–173.

11 Psychology of Organic Food Choice

> Food to a large extent is what holds a society together and eating is closely liked to deep spiritual experiences. Peter Farb and George Armelagos in *Consuming Passions: The Anthropology of Eating*.

The issue of what motivates consumers to purchase organic food has been addressed by many researchers. It is a very interesting question because the growth of the organic food industry can be described as a bottom-up development rather than a top-down one. As a result, much of the information on organic foods has until recently been anecdotal rather than from official agencies or research institutions. It is little wonder, then, that consumers and the food industry have been confused on important aspects of organic food.

The motivation to opt for organic food over conventional food has been ascribed to several issues, as described in Chapter 2, as follows:

(1) *Safety*: the possible presence of harmful chemical and pesticide residues, also the possible presence of food-poisoning organisms. Included in this category is that the food may have been produced from plants or animals using breeding techniques such as cloning and gene modification.

(2) *Nutritional quality and taste*: many consumers believe that food produced on a large-scale is inferior nutritionally to food produced organically. These are the consumers who would prefer to have their food produced locally on small farms and to have the food delivered fresh to market. They regard the fresh food as being tastier and of higher nutritional quality.

(3) *Environmental issues*: many consumers believe that organic farming is better for the environment. This issue includes concerns about the way in which animals are kept on large farms, there being a widespread belief among consumers that modern production methods are cruel; result in an increase in greenhouse-gas production and global warming; and disrupt the interrelationship of plants, crops, farm animals and wildlife.

Organic Production and Food Quality, First Edition. Robert Blair.
© 2012 John Wiley & Sons, Ltd. Published 2012 by John Wiley & Sons, Ltd.

The safety issue

As shown in Chapter 9, the documented findings indicate that conventionally produced foods of domestic origin do not contain harmful residues of pesticides and chemicals. It is true that a small database indicates that, in general, organically produced food contains fewer pesticide residues than conventionally produced food. However, the levels recorded in the conventional food in question are much lower than the levels that can be considered deleterious, and some are found merely at the level of detection. Another point is that, to date, organic foods have not been subjected to the amount of official testing required for conventional foods, making conclusions on residue levels very difficult. But if organic foods carried much less of a risk from dangerous chemicals, would we not expect organic farmers to be much healthier and longer-living than other individuals? As noted in Chapter 9 there is no evidence that they are.

Nutritional quality and taste

As shown in Chapter 10, organic food is fairly similar to regular food in terms of nutritional content. This information should be very reassuring to most consumers. It is true that some differences exist, for instance organically produced soft fruit is likely to be smaller and less moist and to have a more intense flavor than when grown conventionally. This feature gives the organic variety of strawberries, for example, a better keeping quality. Other examples of reported differences in organic and conventional foods are lower yields of organic fruits and vegetables, an increased content of vitamin C and higher antioxidant levels in organic fruits and vegetables, and lower levels of nitrate in organic vegetables. A lower protein content has been reported in organic wheat. Organic meats are less tender than conventional meats but may have a flavor that is preferred by some consumers.

It is not yet clear whether consumers prefer organic food on the basis of a difference in flavor or taste. As outlined in Chapter 3 this question is complicated by the fact that, for example, produce may not be as fresh or at the same stage of maturity when organic produce and conventionally grown produce are compared. Also, labeling of the samples has been shown to result in a higher rating for the produce labeled "organic" (either correctly or falsely). Tests conducted to avoid these problems (Zhao et al., 2007) showed no significant differences in consumer liking or consumer-perceived sensory quality between organically and conventionally grown red loose-leaf lettuce, spinach, arugula, mustard greens, tomatoes, cucumbers, and onions. The only exception was in tomatoes where the conventionally produced tomatoes were rated as having a significantly stronger flavor than the organically produced tomatoes. This was attributed to the conventionally grown tomatoes being slightly riper than the organically grown tomatoes. Another important consumer aspect of organic foods is whether it looks different and whether this influences choice.

The issue of whether the taste of organic food is better than the taste of conventional food has led to an interesting development in Europe, with the setting up of the Ecropolis project under the coordination of the Research Institute of Organic Agriculture (FiBL), Frick, Switzerland. The purpose of the Ecropolis project is to build a database of findings on the taste of organic food, since there is controversy surrounding this issue. The database will be a valuable source of information for scientists, consumers and the food industry. The project has been funded with over €2 million from the Seventh Research Framework Programme (FP7) of the European Union and runs from 2009 to 2011 (www.fibl.org/en/switzerland/research/food-quality-and-safety/ecropolis-en.html).

It is obvious from the published background to the project that the issue of taste and other sensory aspects of organic food are far from clear, and that this important feature has been recognized by a leading organic research center in Europe. For instance the background states:

> Distributors and promoters of organic food claim superior tastes for their products compared to the conventional alternative. This argument however is still subject to a hard debate and thus deserves more scientific evidence. Since repurchases are dependent on the overall liking of a product, and sensory experiences may have an important impact, knowledge about these dimensions is crucial for producers and marketers of organic food to offer products which meet consumer expectations.

The output of the project will be the "first Organic Sensory Information System (OSIS), a multilingual and centrally based data folder for data deposition along with an interface scheme that serves as a basis for data exchange to the benefit of the organic food market (organic associations, producer, processors, retailer, wholesaler as well as consumers)." The plan is for FiBL to coordinate the work of the different project partners in France, Germany, Italy, Poland, Switzerland, and The Netherlands.

This development highlights the uncertainty surrounding the whole issue of the taste of organic food and whether it really tastes better than conventional food. More importantly, the uncertainty is admitted by leaders of the organic farming industry in Europe. The Ecropolis report should be very interesting when it is published in 2011 or 2012!

The Ecropolis group has already published some findings on taste in their report, "Sensory experiences and expectations of organic food" (Stolz *et al.*, 2010). The findings indicate that taste is an important sensory attribute when consumers evaluate organic food, although it is not always the most important sensory property that motivates consumers to purchase organic food. Appearance and odor appear to be the most important sensory attributes when consumers purchase food, while taste and odor are the most important attributes when the food is eaten.

The importance of taste as a criterion for purchasing organic food was found to vary from country to country. In France, quality and taste were an important source of motivation for eating organic products, especially for consumers already accustomed to eating organic food. In Germany, health was the most important buying motive for consumers of organic food, followed by taste. In Italy, the most important buying motives were "naturalness" (34 percent), "health" (31 percent) and "authenticity" (25.5 percent). A "better taste" was a buying motive of only a few consumers (5 percent). In The Netherlands, there was a consistent pattern in consumer motivations: taste, health, and "friendlier for environmental [*sic*] and animal friendly". Consumers of organic products were motivated by hedonistic values rather than altruistic values. In Poland, health and safety were the most important buying motives of organic consumers, in contrast to other European countries, whereas taste was not mentioned at all. In Switzerland, health (37 percent) was an important buying motive, followed by environmental protection and animal welfare. The most important motive for buying organically produced dairy products was animal welfare, followed by personal health. Genuine taste and a preference for organic food in general were also relevant buying motives. Households with a high income bought organic food mainly for altruistic and hedonistic reasons such as environmental awareness and animal welfare as well as for food quality and taste. Households with a lower income primarily bought organic food for animal welfare and health reasons.

Results from focus groups with consumers shed more light on the importance of taste when consumers were asked the following questions:

- Which senses are important to you when eating?
- Did you perceive sensory differences to conventional food when eating organic food?
- On which occasions do you prefer organic or conventional products because of their sensory properties?
- Please take a moment to imagine the sensory experience of organic food. What kind of images come to your mind?
- Basically, do you expect organic products to taste similarly to conventional products, or differently?
- Do you remember situations where your buying decision was influenced by sensory information and how?

The results showed that the following criteria were used by consumers in evaluating sensory properties:

- "Taste" was the most important sensory category for consumers in all countries.
- "Odor" was slightly less relevant in most countries whereas in Italy "odor" had the same importance as taste.
- Consumers often linked the senses of "appearance" and "taste".
- Other senses such as "texture" or "mouth-feel" were of secondary importance for participants.

The results also showed that taste was associated with:

- traditional farming methods;
- childhood memories, "taste of the product as it used to be", which served as a "personal sensory-quality standard";
- a positive image of organic food underlined by a negative image of conventional food.

When asked why sensory attributes in general and especially the taste of organic food should be different compared with conventional food, the responses were:

- Organic food should not be standardized: it should differ from conventional in terms of variability and sensory aspects.
- A stronger, more intrinsic, taste was expected in organic foods, especially in unprocessed commodities such as meat, vegetables, fruit and milk.
- Organic products should have a more "authentic", "healthy", "natural" taste which is "more intense", also with a higher diversity in appearance and taste.
- Processed organic foods such as yogurt, biscuits, sweets should be prepared according to typical organic recipes and have the typical taste of these foods.
- Consumers expect a lower level of certain "unpleasant" constituents in organic foods, such as sweetness, fat and salt. Also they expect a higher amount of constituents such as fruits.
- The taste of organic food ought to be provided by basic ingredients such as whole wheat flour, rather than by secondary ingredients such as additives.
- A different taste can justify higher prices.

Further aspects related to the variability of organic products were given:

- Taste can vary with different points of sale (e.g. supermarket taste expected as conventional food, wholefood-store organic food expected to taste different).
- Despite dissatisfaction with certain sensory aspects of organic food, consumers are tolerant of different taste components: taste is to be learned.

In general the respondents indicated a preference for stronger, more intense (less watery) flavors in all categories of foods (vegetable produce, fruit, dairy products, meat and bakery products).

For many respondents a "less regular appearance" and "less perfect shapes" were regarded as criteria for organic quality and linked with a guarantee of superior taste.

The report also noted that:

> An important word used to describe the differences between organic and conventional food was "authenticity". Concerning taste, consumers showed a strong desire to experience taste as it used to be in the past, a more natural or intense taste. This was in line with the associations and images that organic consumers ascribe to organic food. According to most consumers, organic food is associated with small-scale, handmade and natural production and with peasant farming. Consumers' sensory expectations are often linked with childhood memories. Their memories seem to serve as a "personal sensory-quality standard" when taste experiences of childhood or former times are compared with modern day sensory characteristics of food. Consumers expect that organic products should not be standardized and should differ from conventional products in terms of variability and sensory aspects.

So far the studies conducted by the Ecropolis group have not involved actual taste tests to determine whether consumers in Europe are able to discriminate between organic and conventional foods based on taste.

Environmental issues

This topic has not been dealt with in detail in previous chapters, therefore it is useful to explore its validity in relation to conventional and organic foods. However, it is a large topic and can only be touched on briefly in its relevance to organic food.

As indicated above, one reason why some consumers are motivated to purchase organic food rather than conventional food is the way the food is produced. These are the consumers who are turning away from mass-produced food, preferring food from small farms rather than from "factory farms". Their attitudes range from a simple desire to support local farmers to a strong desire for environmental sustainability. This motivation is very powerful and in some cases borders on the religious. For the consumer who feels this way about food and is prepared to pay the premium for organic food, it is important to buy organic rather than conventional food.

There are many parallels between the organic community worldwide and religion. The organic movement has its leaders (high priests) who set out the principles of organic farming, based on the teachings of mystics such as Rudolph Steiner. The principles are developed from an idealistic rather than a researched approach. Included in the philosophy is the belief that food produced organically possesses benefits not measurable by current scientific methods.

One outcome of this difference in approach is that the aim in the growing of organic food crops is the maintenance of soil fertility rather than the provision of nutrients for plant growth. Environmental sustainability is a main aim of organic production, producers being prepared to accept a 20–45 percent reduction in crop yield to achieve this objective. As a result the organic system is less efficient in food production and requires larger areas of land for its production, but organic producers contend that the produce is better than the produce grown conventionally.

Several authors have reviewed the environmental aspects of organic crop production, mainly from a sustainability aspect since this is an important aim of organic farming. Kirchmann and Thorvaldsson (2000) in their review found that environmental problems in agriculture vary from one country to another. Some of them are caused by natural conditions (such as high native heavy metal content, drought, volcanic eruptions, etc.); others depend on agricultural practices (such as leaching of nutrients and pesticides, etc.); and some are related to human influence in other areas (such as air pollution). Furthermore, these causes are often interrelated. Many environmental problems found with conventional agriculture are also present with organic agriculture. In addressing these problems Kirchmann and Thorvaldsson (2000) observed that several European and other countries focus on organic farming as a solution, but concluded that this approach is dangerous because it does not necessarily lead to a better environment or better food products. Instead they advised that innovative, creative solutions and discoveries based on natural sciences will be helpful in the development of sustainable agriculture, but not methods based on dogmatism.

In a subsequent review Kirchmann and Ryan (2004) reported that expectations about the superiority of organic farming methods with respect to nutrient use efficiency, soil fertility, nitrate leaching and nutrient recycling are not justified by scientific studies. Their review examined the implications of organic farming fertilizer practices for the sustainability of farming systems using two contrasting regions, Europe and Australia. In both these regions, average yields are generally 20–45 percent lower on organic farms than on conventional farms, primarily due to reduced levels of plant-available nutrients. Changes in the population of soil organisms do not overcome this limitation. Nutrient inputs are lower on organic farms, although in Europe there is a tendency on organic farms for increased application of purchased, approved nutrient sources other than fodder. However, these inputs simply allow organic farms to obtain nutrients that originated from conventional farms. If organic farming were to be widely adopted, lower yields would require more land (25–82 percent) to sustain production. In Europe, organic practices increase nitrate leaching, both per unit area and per unit of food produced, due to lower efficiency of nitrogen usage. Despite their aim of maximizing nutrient recycling, organic farming systems recycle only on-farm wastes and approved food wastes. In future, readily soluble inorganic fertilizers will be extracted from organic wastes through new nutrient recovery technologies, and this will make conventional agriculture more sustainable whereas organic farming excludes itself from non-farm recycling, no matter how environmentally clean and safe the new fertilizer products are. In conclusion, the current promotion of organic principles irrespective of environmental outcomes means organic farming has become an aim in itself. This approach is ideological, not scientific, according to Kirchmann and Ryan (2004), and may exclude other more effective solutions to the environmental problems afflicting current agricultural systems.

Another outcome of the difference between idealism and findings based on research is that large sectors of the organic poultry and livestock industry in Europe are having to be curtailed because of a shortage of organic feedstuffs (solvent-extracted oilmeals are not approved as organic feedstuffs). As a result Europe has introduced derogations (exceptions to the rules) that allow a proportion of non-organic feedstuffs to be used until such time as an adequate supply of organic feedstuffs is available. This situation is predicted to be likely to continue for quite some time. The situation has in addition made comparisons of the effects of feeding organic and conventional diets to animals very difficult and of doubtful validity, many of the so-called organic diets not being organic at all. The situation has also had environmental repercussions. Many of the organic feed formulations are not as well balanced nutritionally as conventional diets, since pure forms of the amino acids lysine and methionine are not approved for inclusion in organic diets. As a result, the organic feed mixtures fed to poultry and pigs commonly provide an excessive amount of nitrogen in the form of protein, resulting in an increased excretion of nitrogen into the environment.

Whether the phenomenon of the halo effect extends to environmental issues is debatable, but it is apparent that use of the term "natural" in relation to organic food brings up an image of the harmony of crop and animal production with the environment and of a reduction in the "carbon footprint". Several researchers have investigated these issues in relation to the motivation to purchase organic food. Given the situation it seems reasonable to propose that the motivation is based on perception rather than factual knowledge. Related to this is the production of animals and their contribution to greenhouse gas production. Analysis indicates that organic cattle production is not as benign as it appears initially. As outlined by Blair (2011), a common perception is that traditional pasture-based, low-input dairy and beef systems are more in keeping with environmental stewardship than modern beef and milk production systems. To test this theory Capper *et al.* (2009) compared the environmental impact of US dairy production in 1944 and 2007. They calculated that the carbon "footprint" per billion kilograms of milk produced in 2007 was 37 percent that of equivalent milk production in 1944. Farming methods have also been compared internationally from an environmental aspect. The situation was shown dramatically in a comparison of a modern dairy farm in Wisconsin with one in New Zealand in which the animals grazed extensively (Johnson and Johnson, 1995). Using total farm emissions per kilogram of milk produced as a parameter, the researchers showed that production of methane from belching was higher in the New Zealand farm, whereas carbon dioxide production was higher in the Wisconsin farm. Output of nitrous oxide, a gas with an estimated global warming potential 310 times that of carbon dioxide, was also higher in the New Zealand farm. Methane from manure handling was similar in the two types of farm. The explanation for the finding relates to the different diets used on these farms, being based more completely on forage (and hence more fibrous) in New Zealand and containing less concentrate than in Wisconsin. Fibrous diets promote a higher proportion of acetate in the gut of ruminant animals, resulting in a higher production of methane that has to be released by belching. When cattle are given a diet containing some concentrates (such as corn and soybean meal) in addition to grass and silage, the pattern of ruminal fermentation alters from acetate to mainly propionate. As a result methane production is reduced. Since methane represents a significant loss of dietary energy, reducing methane production in the gut also improves the efficiency of conversion of feed energy into beef or milk.

The environmental impact of animal farming (conventional and organic) is therefore an issue that is currently being reviewed by scientists and legislators as part of the discussions on global warming. One response to the issue is that the European Common Agricultural Policy has been revised to add supplementary measures that include the environmental role of agriculture. The revision includes the adoption of a life-cycle assessment to estimate emissions per kilogram of CO_2 equivalent per kilogram of live weight leaving the farm gate per annum and per hectare. It is possible that in certain countries the legislation will place restrictions on cattle and sheep production, including organic farms. One particular aspect of the environmental issue that is being studied under this program is the influence of organic cattle production on greenhouse gas emissions, particularly methane. This gas is considered to have 21 times the global-warming potential of CO_2. It has been estimated that beef production world-wide accounts for about 62 percent of total livestock methane emissions, milk 19 percent, sheep 12 percent, pigs 5 percent and poultry 1 percent. Estimates suggest that livestock in Asia and the Pacific produce 33 percent of total methane emissions, Latin America 23 percent, Europe 14 percent, Africa 14 percent, North America 11 percent and Oceania 5 percent. As already noted, organic farmers need to avoid low-quality pastures and forages. Natural, unimproved pastures may seem to fit better with organic animal production but these pastures and the forage they support are the ones associated with higher methane emissions. An added benefit in improving the pastures and grazing areas is better growth and milk production, which should lead to increased profits and less of a likelihood that future legislation could lead to a reduction of animals on that farm.

An interesting question is whether consumers will continue to purchase organic foods once they become more familiar with the facts relating to the perceptions outlined here. After all, organic food is more costly. And why is it more costly? Organic food is more costly to produce, which is the main reason it is more expensive at the retail level. Also, there is less of it. The yield of most crops is 20–50 percent lower than the yield of conventional crops, due to differences in soil fertilization and pest control. Organic dairy cows produce less milk. Organic pigs and cattle are more slow-growing and take longer to reach market size. The types of hens suitable for outdoor production lay fewer eggs and they require more feed because of increased energy expended on exercise and in keeping warm. Organic-meat chickens take 2–4 weeks longer to reach market weight than conventional chickens. All of these factors account for the higher cost of organic food.

Image

For consumers who feel strongly about organic food a statement that it is fairly similar to regular food in terms of safety from chemical residues and nutritional content is unacceptable, even when supported by the scientific evidence. The image of organic food is a strong motivating force in their purchase of food and it seems clear that for them the purchase of organic food will continue regardless of the scientific findings on relative safety and nutritional quality. Part of the image is that organic food is perceived as being fresher, particularly if produced on local farms. As noted previously, many consumers share this perception and prefer their food (produce especially) to be fresh and not stored for long periods before purchase.

This very positive effect of the image of organic food was demonstrated in results from a study in Denmark (Scholderer *et al.*, 2004). Each participant (total 185) tasted eight pork-chop (loin) samples from pigs raised either conventionally or organically. The meat was labeled as: (i) organic pork; (ii) free-range pork, i.e. the pigs had access to outdoors but had not been given organic feed; (iii) conventional pork; or (iv) unlabeled. This gave eight samples (2 types × 4 labels, Table 11.1). In other words, the label did not necessarily show the true origin of the meat. This clever approach allowed the researchers to test how expectation affected the overall assessment of quality.

Table 11.1 Ratings (185 participants) for conventional and organic pork meat in a Danish taste test and the willingness of consumers to pay for the pork

	Actual meat type	
	Conventional Average rating	Organic Average rating
Expected quality		
Label info: none	5.08	5.02
Label info: conventional	5.12	4.93
Label info: free range	5.46	5.36
Label info: organic	5.33	5.30
Determined quality: taste		
Label info: none	5.15	4.89
Label info: conventional	4.98	4.85
Label info: free range	5.63	5.58
Label info: organic	5.65	5.50
Determined quality: juiciness		
Label info: none	5.24	5.12
Label info: conventional	4.96	4.69
Label info: free range	5.43	5.22
Label info: organic	5.43	5.11
Determined quality: tenderness		
Label info: none	5.10	4.89
Label info: conventional	4.89	4.88
Label info: free range	5.38	5.33
Label info: organic	5.34	5.11
Determined quality: overall acceptability		
Label info: none	5.24	4.87
Label info: conventional	4.94	4.81
Label info: free range	5.47	5.46
Label info: organic	5.45	5.36
Willingness to pay (DKK per kg)		
Label info: none	88.71	84.13
Label info: conventional	86.28	81.89
Label info: free range	95.88	95.15
Label info: organic	98.04	94.84

Note: ratings were on a scale of 1–7 except for willingness to pay.

(*Source*: Scholderer, J., Nielsen, N.A., Bredahl, L., *et al.* (2004). Organic pork: consumer quality perceptions. Report No. 02/04. Aarhus School of Business, Aarhus, Denmark. Reprinted with kind permission of the authors.)

Before tasting each sample, the participants rated the expected quality. After tasting each sample, the participants rated the quality based on four attributes (taste, tenderness, juiciness, overall acceptability) then stated how much they would be prepared to pay for that sample (willingness to pay).

The results were very enlightening. When asked to rate the pork by its appearance, consumers rated the pork that was labeled organic or free-range higher than pork that was unlabeled or labeled as conventional. This occurred regardless of actual pork type. The effects of label information were calculated to be on average nine to ten times higher than the effects of actual pork type. The consumers were obviously greatly influenced by the information on the label, confirming other findings that the perceived quality of organic meat is largely governed by expectation. However, the organic pork received consistently lower ratings than conventional pork when tasted, regardless of label information. Although the differences were small, they were statistically significant. This result showed that the organic pork tested in this study could not be claimed as being higher in quality than conventional pork, in spite of expectations. The findings also showed that higher ratings were given when the pork was labeled as organic or free-range. Finally, the study showed that consumers were willing to pay more for pork labeled as organic or free-range even though it received lower rankings in the taste test.

This study demonstrated the very powerful effect of image, and how that image influenced the perception of the other qualities of the food.

This effect of the image of organic food is very real, and Linder *et al.* (2010) demonstrated that it is associated with brain activity, namely involvement of the subcortical dopaminergic reward system. In a laboratory study the German scientists used 30 healthy, right-handed participants (15 men and 15 women of mean age 26.03 years) and measured their brain activity while they were shown pictures of organic and conventional foods. The participants were also asked to record how much they were willing to pay for each product as they saw its picture. The subjects saw pictures of 40 different food items (twice each) inside a scanner (80 total items), the foods being a typical variety of foods that were readily available in both organic and conventional forms. The participants were highly familiar with the foods in question. Half of the foods had the German organic emblem, indicating the item had been produced organically. The other half, showing exactly the same food pictures, had an artificially created logo indicating conventional production of that food. Subjects were instructed that the foods with the organic logo were of organic production, and the foods with the other logo were of conventional production. Participants were asked to fast 4 hours before the start of the experiment. Before entering the scanner, each participant received a brochure about the meaning of the organic emblem. The instructions informed subjects that every item shown could be purchased, based on a bidding system.

The results on brain activity showed increased neural activity in the ventral striatum during the presentation of the pictures of organically labeled food, in contrast to the presentation of pictures of conventional food. Associated with this was a willingness by the subjects to offer to pay (bid) 44 percent more for the organic food. The effect produced by the organic foods was attributed to a reward system in the brain, an effect that has been recorded in other studies such as those involving the purchase of cultural goods indicating wealth and status (e.g. sporting and luxury automobiles) and eating desserts and other palatable high-calorie foods that cause reward stimuli.

As noted in Chapter 1, another psychological influence related to the purchase of organic food is the halo effect, which is an additional and important motivating factor. This effect was discovered by Thorndike (1920) in the course of research aimed at testing the ability of supervisors to judge their subordinates on the basis of supposedly independent attributes. He found that judgments on each independent attribute were strongly influenced by the overall impression of the person being judged. In other words the assessments of specific attributes possessed by a person were not arrived at in an unbiased way but were based on an overall feeling about whether the person was rather good or rather inferior. This he called the halo effect.

Researchers studying the purchase of organic food have noted the existence of the halo effect (e.g. Schutz and Lorenz, 1976; Johansson *et al.*, 1999). In both studies, information about the way the food had been produced influenced its acceptability to consumers. When compared with unlabeled foods, the same products labeled as organic generally showed increases in measures of preference. Thus, both studies suggest that consumers have expectations regarding the superior taste of organic produce based on label information. The researchers suggested that such beliefs may be reinforced by repeated consumption of organic produce. One other reason for the popular belief in the flavor superiority of organic produce is that such produce might be consumed in a more optimal state of freshness.

Another example of the halo effect was demonstrated by Schuldt and Schwarz (2010). These researchers tested the responses of university students to web pages displaying nutritional information on conventional cookies or similar cookies "made with organic flour and sugar". The participants were then asked, "Compared to other cookie brands, do you think that 1 serving of these [organic] cookies contains fewer calories or more calories?" and "Compared to other cookie brands, how often should these [organic] cookies be eaten?" The results showed that the students inferred that the organic cookies were lower in calories and could be eaten more often than conventional cookies. This effect was observed even when the nutrition label conveyed an identical calorie content. The effect was more pronounced among participants who were strong supporters of organic production and had strong feelings about environmental issues.

Another group of students was asked to comment on the hypothetical example of a young woman who was on a weight-loss program, and to indicate the decision they thought was best in relation to a slight dilemma she was facing. The dilemma was that she was considering foregoing her usual run of 3 miles that evening to spend more time on schoolwork and whether it was acceptable to eat a dessert after a dinner of roasted vegetables and rice. The researchers found that foregoing exercise was deemed more acceptable by the students when the subject chose organic rather than conventional dessert, even though she was on a weight-loss program. As a result they concluded that the results reflected an "organic/natural"–"healthy" association that was capable of biasing everyday judgments about diet and exercise.

The perception that organic food is low-calorie food or health food appears to be quite common, as noted in Chapter 1.

Conclusions

The motivation to purchase organic food is based on three main perceptions: that it is devoid of the chemical residues found in conventional food; that food produced on a large scale is inferior nutritionally to food produced organically; and that growing organic food is better for the environment. The available evidence in support of these perceptions is weak, except for some aspects of the environmental issue. One factor that drives consumers to purchase organic food is the presence of an organic label, which leads many consumers to infer qualities that are not substantiated factually.

References

Blair, R. (2011). *Nutrition and Feeding of Organic Cattle*. CAB International, Wallingford, Oxford, UK.

Capper, J.L., Cady, R.A. and Bauman, D.E. (2009). The environmental impact of dairy production: 1944 compared with 2007. *Journal of Animal Science* **87**, 2160–2167.

Ecropolis project under the coordination of the Research Institute of Organic Agriculture (FiBL), Frick, Switzerland.

Farb, P. and Armelagos, G. (1980). *Consuming Passions: The Anthropology of Eating*. Houghton Mifflin, Wilmington, Massachusetts.

Johansson, L., Haglund, A., Berglund, L., *et al.* (1999). Preference for tomatoes, affected by sensory attributes and information about growth conditions. *Food Quality and Preference* **10**, 289–298.

Johnson, K.A. and Johnson, D.E. (1995). Methane emissions from cattle. *Journal of Animal Science* **73**, 2483–2492.

Kirchmann, H. and Ryan, M.H. (2004). Nutrients in organic farming – are there advantages from the exclusive use of organic manures and untreated minerals? Proceedings of the 4th International Crop Science Congress, Brisbane, Australia, p. 828.

Kirchmann, H. and Thorvaldsson, G. (2000). Challenging targets for future agriculture. *European Journal of Agronomy* **12**, 145–161.

Linder, N.S., Uhl, G., Fliessbach, K., *et al.* (2010). Organic labeling influences food valuation and choice. *NeuroImage* **53**, 215–220.

Scholderer, J., Nielsen, N.A., Bredahl, L., *et al.* (2004). Organic pork: consumer quality perceptions. Report No. 02/04. Aarhus School of Business, Aarhus, Denmark.

Schuldt, J.P. and Schwarz, N. (2010). The "organic" path to obesity? Organic claims influence calorie judgments and exercise recommendations. *Judgment and Decision Making* **5**, 144–150.

Schutz, H.G. and Lorenz, O.A. (1976). Consumer preferences for vegetables grown under "commercial" and "organic" conditions. *Journal of Food Science* **41**, 70–73.

Stolz, H., Jahrl, I., Baumgart, L. and Schneider, F. (2010). Sensory Experiences and Expectations of Organic Food. Results of Focus Group Discussions. Research Institute of Organic Agriculture (FiBL), Frick, Switzerland, and Frankfurt, Germany.

Thorndike, E.L. (1920). A constant error in psychological ratings. *Journal of Applied Psychology* **4**, 469–477.

Zhao, X., Chambers, E., Matta, Z., *et al.* (2007). Consumer sensory analysis of organically and conventionally grown vegetables. *Journal of Food Science* **72**, S87–S91.

12 Conclusions

This review of the published findings concludes that organic and conventional foods are fairly similar in terms of their nutritional quality and freedom from harmful chemical residues. That conclusion is in agreement with conclusions reached by many other scientists and governmental food agencies worldwide.

A very surprising development is that an important group within the organic industry in Europe now agrees with this assessment. As outlined in Chapter 11, the Ecropolis project has been established under the coordination of the Research Institute of Organic Agriculture (FiBL), Frick, Switzerland, its purpose being to build a database of findings on the taste of organic food since this is a controversial issue.

The background documentation related to the establishment of Ecropolis includes the following statement (Canavari *et al.*, 2009):

> there is broad agreement on two points:
>
> - there is no proof that organic food is more nutritious or safer, and
> - most studies that have compared the taste and organoleptic quality of organic and conventional foods report no consistent or significant differences between organic and conventional produce. Therefore, claiming that all organic food tastes different from all conventional food would not be correct. However, among the well-designed studies with respect to fruits and vegetables that have found differences, the vast majority favour organic produce. Organic produce tends to store better and has longer shelf life, probably because of lower levels of nitrates and higher average levels of antioxidants. The former can accelerate food spoilage, while antioxidants help preserve the integrity of cells and some are natural antibiotics.

Given that surprising agreement, it is understandable that the decision was made to concentrate the Ecropolis project on the next perceived attribute of organic food: taste. The results obtained by the Ecropolis project should be very interesting and it is to be hoped that the scientists involved will be able to separate the effects of taste and freshness. It seems clear that

---

Organic Production and Food Quality, First Edition. Robert Blair.
© 2012 John Wiley & Sons, Ltd. Published 2012 by John Wiley & Sons, Ltd.

much of the appeal of organic food, especially when purchased at farmers' markets, is its freshness. It may well be that freshness emerges as a more important attribute in its effect on taste than taste *per se*. More research needs to be conducted on this issue since freshness is an important attribute for the food shopper. This change demonstrates the important influence the organic industry has had on the appreciation of food quality by the public, and no doubt will have an important influence on the conventional food industry.

It is true that some differences have been recorded between organic and conventional food. Some studies have shown an increased content of vitamin C, β-carotene and phenolic compounds in organic produce, but a study with humans found that a high intake of these compounds from tomatoes did not lead to an increase in plasma concentration. Some vegetable produce has a lower nitrate content when grown organically. The significance of dietary nitrate content and that of phenolic compounds in relation to human health has not yet been established. As with vegetable produce, there is evidence that some organic fruit is drier than convention- ally grown fruit but is similar in nutrient content to conventional produce on a moisture-corrected basis. A slightly drier fruit is likely to have a more intense flavor and as a result may be preferred by the consumer. A drier fruit is also likely to have a higher resistance to deterioration and better keeping quality. Higher contents of phenolic compounds and antioxidants have been reported in some organic fruit, the significance of which in terms of human health is not yet clear.

Organic cereals generally have a lower content of protein, but in other nutritional aspects are similar to conventional cereals except for a slightly higher content of phenolic compounds. Animals appear to have the ability to discriminate between organic and conventional sources of cereals. In preference tests, rats selected biscuits made from organic wheat over biscuits made from conventional wheat. Conversely, wild birds preferred conventional wheat seed as a winter feed, a result ascribed to a lower content of protein in the organic wheat.

Beef animals raised organically grow more slowly and produce leaner carcasses. As a result the meat tends to have less marbling and is less tender. The profile of the fat is altered with organic production (or with grass feeding), with a higher content of PUFAs (in particular CLA). Similar findings have been reported with pigs and poultry, the research and consumer findings suggesting that the result is meat that is slightly tougher but has an enhanced flavor that is preferred by some consumers (probably an age effect since the organic animals take longer to reach market weight). The main difference between organic (farm-raised) and wild fish is a higher content of fat with a higher content of PUFAs in the organic fish.

Organic production, or at least forage consumption, is known to result in a higher content of PUFAs in milk, mainly in α-linolenic acid. Organic milk may also taste differently from regular milk, especially if it has been heat-treated by UHT to extend its keeping quality. There is no evidence that the cholesterol content of eggs is lower in organic than in regular eggs, although the fatty acid profile may be more favorable. The documented evidence indicates that organic and conventional eggs are otherwise similar in nutritional characteristics, although organic eggs may be smaller and may be brown-shelled rather than white-shelled (a breed effect).

All of the findings indicate that the consumer who buys food strictly on price can be assured that conventional food is just as safe and of the same nutritional quality as organic food. The consumer who demands freshness in the food, has strong feelings about how food is produced, feels that organic food is more "healthful" and is willing to pay a premium should be encouraged to buy organic food. The benefits are tangible, although they may not be quite what they are assumed to be. The organic label on food has a very powerful effect on some consumers, resulting in a halo effect and expectations that cannot be supported factually.

It is not surprising that consumers should have the perception that organic food is superior to conventional food, since most of the public relies on the media for information. Dan Gardner, writing in the *Ottawa Citizen* on November 22, 2008, wrote an interesting article on perception of the news in the media (reprinted with permission).

Consider this headline in the New York Times: "As crop prices fall, farmers face losses." Or this one from the Canadian Press: "Plunging commodity prices hurt miners." And you know all those dire stories that appeared this week warning that we may experience deflation? Six months ago, we were warned about inflation. Or worse, stagflation. The fact that deflation is a serious possibility now means those threats have vanished. But you don't put that in the headlines because good news isn't news.

News, by definition, is bad.

They don't actually teach that in journalism school, and I've never heard any editor put it quite that way, and yet it is a fundamental truth about the news. Soaring commodity prices? Disaster looms. Falling commodity prices? Disaster looms. Whatever happens, folks, it's bad. Real bad.

Here's an interesting little fact: Over the last 15 years, child mortality rates in China fell by 50 per cent. As a result, millions of children have been saved from a tragic death.

Ever read that anywhere? Probably not. I've been giving lectures on risk perception in Canada and elsewhere and I always ask the audience if they have heard that. Out of the couple of thousand people I've quizzed, roughly five have said yes.

I follow that question with another: And how many of you have heard of Madeleine McCann, the British toddler who was abducted and presumably murdered in Portugal? Always, every hand in the audience goes up.

As a reflection of reality, popular perception is weirdly distorted. As a reflection of the media, it's as clear and accurate as a well-polished mirror.

At the other end of the spectrum of information on organic food are the gushing announcements and exaggerated claims on the internet, prime examples of the power of the halo effect.

The basis for the perception of the superiority of organic foods by the public is not, therefore, difficult to find.

The admission by a leading group within the organic farming industry that there is now broad agreement that organic and conventional foods are similar in nutritional value and in safety from harmful residues (Canavari et al., 2009) should be taken as a signal that the acrimonious debates that have taken place on this issue need to end. The issue is settled. Other points that support this view are that life expectancy continues to increase, not decrease. Also, organic farmers are no healthier than other members of the public, a result opposite to what would be expected if the diet enjoyed by the average consumer was harmful.

It would be more meaningful to redirect research efforts to areas that merit further investigation, such as taste and freshness. Among these are the issues of nitrates and phenolic contents in foods, which are known to differ in foods produced organically and conventionally. What is the significance of these findings in relation to human health? And do the findings help to explain the higher life expectancy found in some Asiatic countries?

The higher dry-matter content of some organically grown soft fruit is an important finding in that it results in fruit with a more intense color and flavor and improved keeping quality. This finding needs to be investigated further with a view to extending these benefits to other fruits. Meat, milk and eggs can be produced with an enhanced content of PUFAs by feeding the animals and birds organic feed or feed containing forage. What is the significance of this finding in relation to human health?

The impact on the environment of organic food production is not as benign as it appears at first, and merits further detailed study based on scientific findings rather than on untested principles. This approach of a more proven basis as opposed to a principle-based set of guidelines for food production could be usefully extended to other areas of the organic industry. An outcome of the conflict between idealism and practicality is that the production of organic meat, milk and eggs in Europe is having to be curtailed because of a shortage of organic feedstuffs, and partly organic diets are having to be allowed. The situation is such that sources such as insect larvae and seaweed are being researched as organic feedstuffs. The refusal to allow solvent-extracted soybean meal in favor of expeller soybean meal only hurts the organic farmer by exacerbating the shortage of organic feedstuffs. On what basis can the refusal be justified? Also, why cannot the pure forms of amino acids be allowed in organic diets for animals to help to rectify the feed shortage situation (as many in the organic industry would support)? After all, synthetic vitamins are allowed as well as minerals such as calcium and phosphorus sources. Also allowed is potato protein concentrate as a feedstuff although this is essentially an industrial by-product and furthermore may have been derived from GM potatoes.

It is to be hoped that this book will benefit consumers by encouraging the media and the food industry to present a more accurate picture of the relative quality of conventional and organic foods, and will benefit researchers and the food industry in its presentation of the up-to-date facts on organic foods and its suggestions for strengthening the important organic sector of the food industry.

Reference

Canavari, M., Asioli, D., Bendini, A., *et al.* (2009). Summary Report on Sensory-related Socio-economic and Sensory Science Literature about Organic Food Products (2009). http://orgprints.org/17208/2/deliverable_1_2_sensory_literature.pdf, accessed October 12, 2010.

Appendix

Concentration measures and equivalents

1 part per million (ppm)
= 1 milligram per kilogram (mg/kg)
= 1 milligram per liter (mg/liter)
= 1 microgram per gram (µg/g)

1 part per billion (ppb)
= 1 microgram per kilogram (µg/kg)
= 1 microgram per liter (µg/liter)
= 1 nanogram per gram (ng/g)
= about three seconds out of a century

1 part per trillion (ppt)
= 1 picogram per gram (pg/kg or µg/kg)
= about three seconds out of one hundred thousand years

Organic Production and Food Quality, First Edition. Robert Blair.
© 2012 John Wiley & Sons, Ltd. Published 2012 by John Wiley & Sons, Ltd.

Index

Organic Production and Food Quality, First Edition. Robert Blair.
© 2012 John Wiley & Sons, Ltd. Published 2012 by John Wiley & Sons, Ltd.

Food Science and Technology

GENERAL FOOD SCIENCE & TECHNOLOGY, ENGINEERING AND PROCESSING

Organic Production and Food Quality: A Down to Earth Analysis	Blair	9780813812175
Handbook of Vegetables and Vegetable Processing	Sinha	9780813815411
Nonthermal Processing Technologies for Food	Zhang	9780813816685
Thermal Procesing of Foods: Control and Automation	Sandeep	9780813810072
Innovative Food Processing Technologies	Knoerzer	9780813817545
Handbook of Lean Manufacturing in the Food Industry	Dudbridge	9781405183673
Intelligent Agrifood Networks and Chains	Bourlakis	9781405182997
Practical Food Rheology	Norton	9781405199780
Food Flavour Technology, 2nd edition	Taylor	9781405185431
Food Mixing: Principles and Applications	Cullen	9781405177542
Confectionery and Chocolate Engineering	Mohos	9781405194709
Industrial Chocolate Manufacture and Use, 4th edition	Beckett	9781405139496
Chocolate Science and Technology	Afoakwa	9781405199063
Essentials of Thermal Processing	Tucker	9781405190589
Calorimetry in Food Processing: Analysis and Design of Food Systems	Kaletunç	9780813814834
Fruit and Vegetable Phytochemicals	de la Rosa	9780813803203
Water Properties in Food, Health, Pharma and Biological Systems	Reid	9780813812731
Food Science and Technology (textbook)	Campbell-Platt	9780632064212
IFIS Dictionary of Food Science and Technology, 2nd edition	IFIS	9781405187404
Drying Technologies in Food Processing	Chen	9781405157636
Biotechnology in Flavor Production	Havkin-Frenkel	9781405156493
Frozen Food Science and Technology	Evans	9781405154789
Sustainability in the Food Industry	Baldwin	9780813808468
Kosher Food Production, 2nd edition	Blech	9780813820934

FUNCTIONAL FOODS, NUTRACEUTICALS & HEALTH

Functional Foods, Nutraceuticals and Degenerative Disease Prevention	Paliyath	9780813824536
Nondigestible Carbohydrates and Digestive Health	Paeschke	9780813817620
Bioactive Proteins and Peptides as Functional Foods and Nutraceuticals	Mine	9780813813110
Probiotics and Health Claims	Kneifel	9781405194914
Functional Food Product Development	Smith	9781405178761
Nutraceuticals, Glycemic Health and Type 2 Diabetes	Pasupuleti	9780813829333
Nutrigenomics and Proteomics in Health and Disease	Mine	9780813800332
Prebiotics and Probiotics Handbook, 2nd edition	Jardine	9781905224524
Whey Processing, Functionality and Health Benefits	Onwulata	9780813809038
Weight Control and Slimming Ingredients in Food Technology	Cho	9780813813233

INGREDIENTS

Hydrocolloids in Food Processing	Laaman	9780813820767
Natural Food Flavors and Colorants	Attokaran	9780813821108
Handbook of Vanilla Science and Technology	Havkin-Frenkel	9781405193252
Enzymes in Food Technology, 2nd edition	Whitehurst	9781405183666
Food Stabilisers, Thickeners and Gelling Agents	Imeson	9781405132671
Glucose Syrups – Technology and Applications	Hull	9781405175562
Dictionary of Flavors, 2nd edition	De Rovira	9780813821351
Vegetable Oils in Food Technology, 2nd edition	Gunstone	9781444332681
Oils and Fats in the Food Industry	Gunstone	9781405171212
Fish Oils	Rossell	9781905224630
Food Colours Handbook	Emerton	9781905224449
Sweeteners Handbook	Wilson	9781905224425
Sweeteners and Sugar Alternatives in Food Technology	Mitchell	9781405134347

FOOD SAFETY, QUALITY AND MICROBIOLOGY

Food Safety for the 21st Century	Wallace	9781405189118
The Microbiology of Safe Food, 2nd edition	Forsythe	9781405140058
Analysis of Endocrine Disrupting Compounds in Food	Nollet	9780813818160
Microbial Safety of Fresh Produce	Fan	9780813804163
Biotechnology of Lactic Acid Bacteria: Novel Applications	Mozzi	9780813815831
HACCP and ISO 22000 – Application to Foods of Animal Origin	Arvanitoyannis	9781405153669
Food Microbiology: An Introduction, 2nd edition	Montville	9781405189132
Management of Food Allergens	Coutts	9781405167581
Campylobacter	Bell	9781405156288
Bioactive Compounds in Foods	Gilbert	9781405158756
Color Atlas of Postharvest Quality of Fruits and Vegetables	Nunes	9780813817521
Microbiological Safety of Food in Health Care Settings	Lund	9781405122207
Food Biodeterioration and Preservation	Tucker	9781405154178
Phycotoxins	Botana	9780813827001
Advances in Food Diagnostics	Nollet	9780813822211
Advances in Thermal and Non-Thermal Food Preservation	Tewari	9780813829685

For further details and ordering information, please visit www.wiley.com/go/food

Food Science and Technology from Wiley-Blackwell

SENSORY SCIENCE, CONSUMER RESEARCH & NEW PRODUCT DEVELOPMENT

Title	Author	ISBN
Sensory Evaluation: A Practical Handbook	Kemp	9781405162104
Statistical Methods for Food Science	Bower	9781405167642
Concept Research in Food Product Design and Development	Moskowitz	9780813824246
Sensory and Consumer Research in Food Product Design and Development	Moskowitz	9780813816326
Sensory Discrimination Tests and Measurements	Bi	9780813811116
Accelerating New Food Product Design and Development	Beckley	9780813808093
Handbook of Organic and Fair Trade Food Marketing	Wright	9781405150583
Multivariate and Probabilistic Analyses of Sensory Science Problems	Meullenet	9780813801780

FOOD LAWS & REGULATIONS

Title	Author	ISBN
The BRC Global Standard for Food Safety: A Guide to a Successful Audit	Kill	9781405157964
Food Labeling Compliance Review, 4th edition	Summers	9780813821818
Guide to Food Laws and Regulations	Curtis	9780813819464
Regulation of Functional Foods and Nutraceuticals	Hasler	9780813811772

DAIRY FOODS

Title	Author	ISBN
Dairy Ingredients for Food Processing	Chandan	9780813817460
Processed Cheeses and Analogues	Tamime	9781405186421
Technology of Cheesemaking, 2nd edition	Law	9781405182980
Dairy Fats and Related Products	Tamime	9781405150903
Bioactive Components in Milk and Dairy Products	Park	9780813819822
Milk Processing and Quality Management	Tamime	9781405145305
Dairy Powders and Concentrated Products	Tamime	9781405157643
Cleaning-in-Place: Dairy, Food and Beverage Operations	Tamime	9781405155038
Advanced Dairy Science and Technology	Britz	9781405136181
Dairy Processing and Quality Assurance	Chandan	9780813827568
Structure of Dairy Products	Tamime	9781405129756
Brined Cheeses	Tamime	9781405124607
Fermented Milks	Tamime	9780632064588
Manufacturing Yogurt and Fermented Milks	Chandan	9780813823041
Handbook of Milk of Non-Bovine Mammals	Park	9780813820514
Probiotic Dairy Products	Tamime	9781405121248

SEAFOOD, MEAT AND POULTRY

Title	Author	ISBN
Handbook of Seafood Quality, Safety and Health Applications	Alasalvar	9781405180702
Fish Canning Handbook	Bratt	9781405180993
Fish Processing – Sustainability and New Opportunities	Hall	9781405190473
Fishery Products: Quality, safety and authenticity	Rehbein	9781405141628
Thermal Processing for Ready-to-Eat Meat Products	Knipe	9780813801483
Handbook of Meat Processing	Toldra	9780813821825
Handbook of Meat, Poultry and Seafood Quality	Nollet	9780813824468

BAKERY & CEREALS

Title	Author	ISBN
Whole Grains and Health	Marquart	9780813807775
Gluten-Free Food Science and Technology	Gallagher	9781405159159
Baked Products – Science, Technology and Practice	Cauvain	9781405127028
Bakery Products: Science and Technology	Hui	9780813801872
Bakery Food Manufacture and Quality, 2nd edition	Cauvain	9781405176132

BEVERAGES & FERMENTED FOODS/BEVERAGES

Title	Author	ISBN
Technology of Bottled Water, 3rd edition	Dege	9781405199322
Wine Flavour Chemistry, 2nd edition	Bakker	9781444330427
Wine Quality: Tasting and Selection	Grainger	9781405113663
Beverage Industry Microfiltration	Starbard	9780813812717
Handbook of Fermented Meat and Poultry	Toldra	9780813814773
Microbiology and Technology of Fermented Foods	Hutkins	9780813800189
Carbonated Soft Drinks	Steen	9781405134354
Brewing Yeast and Fermentation	Boulton	9781405152686
Food, Fermentation and Micro-organisms	Bamforth	9780632059874
Wine Production	Grainger	9781405113656
Chemistry and Technology of Soft Drinks and Fruit Juices, 2nd edition	Ashurst	9781405122863

PACKAGING

Title	Author	ISBN
Food and Beverage Packaging Technology, 2nd edition	Coles	9781405189101
Food Packaging Engineering	Morris	9780813814797
Modified Atmosphere Packaging for Fresh-Cut Fruits and Vegetables	Brody	9780813812748
Packaging Research in Food Product Design and Development	Moskowitz	9780813812229
Packaging for Nonthermal Processing of Food	Han	9780813819440
Packaging Closures and Sealing Systems	Theobald	9781841273372
Modified Atmospheric Processing and Packaging of Fish	Otwell	9780813807683
Paper and Paperboard Packaging Technology	Kirwan	9781405125031

For further details and ordering information, please visit www.wiley.com/go/food